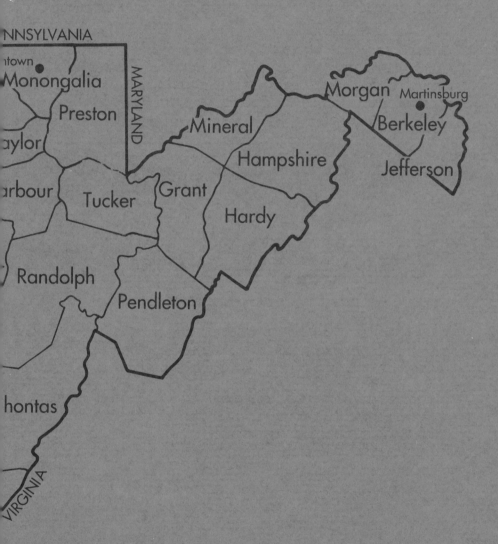

NNSYLVANIA

ntown

Monongalia

Preston

MARYLAND

Mineral

Morgan   Martinsburg

Berkeley

aylor

Hampshire

Jefferson

rbour

Tucker

Grant

Hardy

Randolph

Pendleton

hontas

VIRGINIA

UNIVERSITY OF PITTSBURGH PRESS
in cooperation with the
WEST VIRGINIA DIVISION OF NATURAL RESOURCES
NONGAME WILDLIFE PROGRAM
and
THE BROOKS BIRD CLUB, INC.

Pittsburgh and London

The

West Virginia Breeding Bird Atlas

« »

Albert R. Buckelew, Jr.
and
George A. Hall

Published by the University of Pittsburgh Press, Pittsburgh, Pa. 15260
Copyright © 1994, University of Pittsburgh Press
All rights reserved
Manufactured in the United States of America
Printed on acid-free paper

Library of Congress Cataloging-in-Publication Data

Buckelew, Albert R.
    The West Virginia breeding bird atlas / Albert R. Buckelew, Jr.
and George A. Hall.
        p.   cm.—(Pitt series in nature and natural history)
    Includes bibliographical references (p.   ) and index.
    ISBN 0-8229-3850-2 (cl : acid-free paper)
    1. Birds—West Virginia.   2. Birds—West Virginia—Geographical
distribution   3. Birds—West Virginia—Geographical distribution—Maps.
I. Hall, George A.   II. Title.   III. Series.
QL684.W4B83   1994
598.29754—dc20   94-10162
                                                                        CIP

A CIP catalogue record for this book is available from the British Library.

Eurospan, London

To all the West Virginia Breeding Bird Atlas Project workers
whose contributions made this book possible.

Contents

<< >>

Acknowledgments

<< >>

The West Virginia Breeding Bird Project was a special project of the Brooks Bird Club in cooperation with the West Virginia Division of Natural Resources' (DNR) Nongame Wildlife Program, which provided financial support with funds derived in part from the voluntary contributions of citizens through a state income tax checkoff. Publication of the *Atlas* was made possible by a grant from the National Fish and Wildlife Foundation.

Special thanks are extended to Daniel L. Rice of the Ohio Department of Natural Resources; Robert C. Whitmore of the Division of Forestry, West Virginia University; Chandler S. Robbins of the U.S. Fish and Wildlife Service; and Douglas P. Kibbe of the Vermont Breeding Bird Atlas for their advice, and to Sam Droege of the U.S. Fish and Wildlife Service for reading the manuscript of this book. Sam Droege and the U.S. Fish and Wildlife Service's Patuxent Wildlife Research Center provided Breeding Bird Survey data. Sue Ridd, Daniel W. Brauning, Daniel L. Rice, and Chandler S. Robbins provided preliminary atlas maps for Virginia, Pennsylvania, Ohio, and Maryland.

J. Scott Butterworth, James M. Crum, Karen G. Eye, Kenneth B. Knight, Kathleen C. Leo, Lucille E. Licwov, Robert L. Miles, James C. Pack, Joseph C. Rieffenberger, and Peter E. Zurbuch of the West Virginia DNR assisted the Atlas Project, and Thomas J. Allen, also of the DNR, furnished the cover painting. Drawings from George Miksch Sutton's *An Introduction to the Birds of Pennsylvania* (J. Horace McFarland Co., 1928) are used with the permission of his sister, Dorothy S. Fuller. Ray R. Hicks, Jr. provided the forest cover map. Figures 3, 4, 6, 8, and 9 were adapted from E. L. Core's *Vegetation of West Virginia* (McClain Printing Co., 1966) and were used with permission. Figure 5 was adapted from W. H. Gillespie and E. C. Murriner's *Forest Resources of West Virginia, 1990* (West Virginia Division of Forestry), courtesy of W. H. Gillespie. Computer services were organized and supervised by Walter Kordek, Terrence M. Hingtgen, Christine L. Hudgins, and Craig W. Stihler of the West Virginia DNR and Justin B. G. Skywatcher of Bethany College. Daniel W. Brauning of the Philadelphia Academy of Natural Sciences provided invaluable advice on computer software. K. Diane Cash, Karen D. Currence, Angela M. Hinkle, Alice M. Pingley, Susan K. Stimpson, Leisa M. Teeters, Brenda E. Teter, and Janet D. White assisted with data processing. Alice E. Bowers, Cynthia L. Cecchini, Eleanore M. Hunter, and Kristen L. Cox provided secretarial assistance. Dana J. Garner provided photographic prints, and Irma Jean Counselman provided bibliographical assistance.

Special appreciation goes to Nancy Fleming, who edited the volume, and to the staff of the University of Pittsburgh Press for their assistance in preparing the *West Virginia Breeding Bird Atlas* for publication.

District Coordinators

The West Virginia Breeding Bird Atlas Project could not have been accomplished without the dedication of the district coordinators:

William H. Armstrong (Parkersburg District)
Albert R. Buckelew Jr. (Northern Panhandle District)
Robert S. Dean (Eastern Panhandle Eastern District)
Thomas R. Fox (Charleston District)
Harriet S. Gilbert (Eastern Panhandle Western District)

Acknowledgments

LeJay Graffious (Morgantown District)
George F. Hurley (Charleston District)
Thomas Igou (Huntington District)
Larry B. McArthur (Western Mountains District)
Donald D. Nemanich (Northern Panhandle District)
James D. Phillips (Princeton District)
Carolyn C. Ruddle (Eastern Mountains District)
Ivan R. Schwab (Morgantown District)
Harriet M. Sheetz (Eastern Panhandle Western District)
Craig W. Stihler (Western Mountains District)
Franklin W. Sturgis (Eastern Panhandle Eastern District)
Gerald R. Wilcox (Eastern Panhandle Western District)
Gary Worthington (Beckley District)

Robert Whitcomb coordinated block busting for eastern parts of the state.

Volunteers

More than 300 volunteers reported breeding species to district coordinators.

Terena Adkins
Helen Agee
James Agee
Billie Altemus
Don Altemus
Leslie Anderson
Phyliss Argabrite
Wendell Argabrite
William Armstrong
Ray Ashworth
Lucy Atkins
Sally N. Baer
Richard Banvard
Ted Banvard
Shirley Barker
Lynn Barnhart
Rodney Bartgis
Bill Beatty
Suzanne Behlendorf
Bob Belding
Janet Belding
Janice Bell
Ralph Bell
William Belton
Sarah Bennett
Terry Blake
David Blockstein
John Blomberg
Julia Bird
Joann Bishop
Cathy Blumenfeld
Steve Bolar
Chris Bone
Lela F. Bonner
Elizabeth Bradley
Willa Bragg

William R. Branyan
George Breiding
Dorothy Broemsen
Sandra L. Brown
Albert R. Buckelew, Jr.
Dana A. Buckelew
Sara K. Buckelew
Robert Burrell
Eleanor Bush
Kyle Bush
William Butler
Ruth Cahn
Dale Cain
Ramona Caincross
Ronald Canterbury
James Carter
Betsy Chadwick
Jack Christian
Frances L. Clancy
John Clancy
Jeannie Clark
Crystal Coffman
Carolyn Conrad
Dorothy Conrad
Helen Conrad
George Constantz
Kim Corbin
Stan Corwin-Roach
Betty Crabtree
Kenneth Cross
Jim Crum
Pat Cummins
Nina Dameron
Marlene Daniels
Larry Davis
Deanna Dawson

Robert S. Dean
Tim Dean
Juanita Delancey
David Dendler
Delores DeVaul
Gene Dobbins
Nancy Dobbins
Bobbie Dotson
Tom Dotson
B. A. Dowell
Sam Droege
Greg Eddy
Norval Eddy
Sue Edmonds
Scott Eggerud
Geneva Eister
Cindy Ellis
Jeannine Elliston
Charles Emerson
Kyle L. Emerson
Jeannette Esker
Richard Esker
Jim Evans
Cynthia Ewing
Sissy Feller
John Findley
Kathleen Finnegan
Michael Finnegan
Janet Fletcher
Dawn A. Fox
Thomas R. Fox
Julian Frechtman
Ruth Frechtman
Steve Fretwell
Phillip Gainer
Mildred Gerling

Acknowledgments

Florence Giffin
Harriet Gilbert
Barbara Ann Gillespie
William Glass
Hullet C. Good
Bob Gordon
LeJay Graffious
Emily Grafton
Mark Graham
Mike Griffith
Irenee Groleau
George A. Grubb, Jr.
Curtis Haines
Kathleen Hale
George A. Hall
Ann Harris
Donna Hartley
Carole Hartman
Debra Hausrath
Greg Henger
Wilbur Hershberger
William Hertig
Richard Hogg
Russell Hogg
Carrie Lynn Hoke
Donna Hollingsworth
David Holmes
Dennis Houmard
Susan Houmard
Janet Howe
Marshall Howe
Marshall Howell
Jeanne Hubbard
John Hubbard
Anne Hurley
George Hurley
Eugene Hutton
Reba Hutton
Helen Igo
William K. Igo
John Igou
Matt Igou
Tom Igou
Helen Imbrock
Phil Jackson
Sydney Jacobs
Clark Jeschke
Kay Johnson
Oliver Johnson
Virginia Johnson
Elizabeth N. Johnston
Douglas Jolley
Armel Jones
Eric Jones
Hilarie Jones

John Jones
Steve Jones
Paul Jung
Scott Jung
Carl Karickhoff
Bob Keedy
Mary Keedy
Retha Keene
Douglas Kibbe
Ben Kiff
Lloyd Kiff
Maxine Kiff
Sara Kiner
Gizella Kish
Robert Kletzly
Ken Knight
George Koch
Don Kodak
Walter Kordek
Francis J. Kosowicz
Nevada Laitsch
Jim Lakiotes
Genevieve Lambert
Jackie Laplante
Ralph Lawhorn
E. A. Leavitt
James B. Lee
Kathleen C. Leo
Cam Lewis
Norma Lewis
Letty Limbach
Joan Lowry
Charles Mack
Lillian Mack
Chuck Mapes
Ben Markell
Mary Markusic
Elwood Martin
Rodger McCorrick
Brian McDonald
Janet McMillen
Ann McRae
Grady McRae
Marian Means
Ray Menendez
Susan Meszaros
Larry Metheny
Janet Meyer
Jim Meyer
Caren Miller
James Miller
Michael H. Mills
Donna Mitchell
Don Morton
Marilyn Morton

Robert W. Moss
David Mozurkewich
Betty Mullins
Robert Mullins
Norma Murray
Phillip Murray
William Murray
Janice Musser
Don Nemanich
George Nichols
Sam Norris
Bruce Nots
Howard Ogden
Toni Ogden
Ephe Olliver
Mark Otto
Jesse Oyster
Jim Pack
John Palavido
O. R. Parker
Loretta Patterson
Bart Paxton
Daniel Perry
John Peters
Charles Pierce
Ivarean Pierce
James Phillips
Judy Phillips
Sharon Phillips
Barry K. Pitts
Mike Pizzino
Gregory Pluth
Lynn Pollard
Mark Poppendeck
Donald M. Post
William Pratt
Valerie C. Presley
Roy E. Pruett
Ann Pyle
Roberta Quattlebaum
Cynthia Rank
Paul Rank
Gary Rankin
Bob Reed
Janet Reynolds
Joe Rieffenberger
Mary Moore Rieffenberger
Bob Rine
Chandler S. Robbins
Richard Rogers
Lorraine Rollefson
Georganna Romano
Bill Roody
Carolyn Ruddle
Art Ryan

Acknowledgments

John Schmitt
Ivan Schwab
Chester M. Shaffer
Paul Shaw
Harriet Sheetz
Ada Shipman
Hallie Sims
Harry E. Slack III
Carl Slater
Juanita Slater
Edwin Lee Smith
William Smith
Sharon Spencer
Ruby Sponaugle
Sally Stebbins
Geneva Steele
Roger Stevens
Craig Stihler
Gary Strawn
Ruth C. Strosnider
Frank Sturgis
Cecil Sturm
Beth Taylor
George Taylor
Jesse Taylor

Fred Temple
Pat Temple
Richard L. Terry
Maxine Thacker
Bob Thompson
Florence Thompson
Arline Thorn
William Tolin
George Trimble
Richard Trimble
Mark Turley
Fritz Turner
Mary Twigg
Norma Jean Venable
Gordon D. Vujevic
Karen Vuranch
Gilbert Walker
Merileen Walsh
Judy Ward
George Warrick
Dale Watson
Elizabeth Watson
Harry Watson
Betty Weimer
Neil Weinberg

Claude Wellman
Judith Whitcomb
Robert Whitcomb
Ronald Wigal
David Wilcove
Gerald Wilcox
Edna Wilfong
Claude Williams
Delores Wilson
Don Wilson
Leon Wilson
Rob Wilson
Steven Wilson
Doug Wood
Patricia Wood
Joan Workman
Gary Worthington
Gene Worthington
Gene Wrzosek
Helen Wylie
William Wylie
Judy Yandoh
Ruth Yarrow
Ava C. Zeitz
Elizabeth Zimmerman

THE WEST VIRGINIA BREEDING BIRD ATLAS

<< >>

Introduction

《 》

The West Virginia Breeding Bird Atlas is based on a tradition of mapping the distribution of plant and animal species, a tradition almost as old as the actual study of such species. Various mapping conventions have been used—some maps plotted actual specimen localities, whereas others, mostly by guesswork, filled in the ranges as solid blocks. Many early species maps plotted cumulative historical data and took no notice of the dynamic character of most bird ranges.

The "atlas" technique of mapping the range on a very fine scale confined to a definite time period is a product of the late twentieth century. The system was first devised by plant scientists in Europe, and the application of the idea to the mapping of the breeding ranges of birds was undertaken in Great Britain and some other European countries, as well as Australia. The first project completed was the *Atlas of the Breeding Birds in Britain and Ireland*, published in 1976 (Sharrock 1976). The idea came to North America in the 1970s, and a small-scale project was carried out in Montgomery County, Maryland, under the direction of C. S. Robbins. Atlases in the United States and Canada have concentrated on the state or province level. Vermont (Laughlin and Kibbee 1985) and Maine (Adamus n.d.) were the first states to complete atlas projects, and since that time, most eastern and midwestern states and the eastern Canadian provinces have undertaken such projects. At this writing, several atlases have been published and several more are either in press or in the final stages of preparation.

West Virginia's Atlas Project

The purpose of the West Virginia Breeding Bird Atlas Project was to inventory and plot the distribution of the breeding bird species of the state in a definite time period, 1984 through 1989. A publication by G. A. Hall, *West Virginia Birds* (1983), had outlined the distribution, mostly on a county-by-county basis, as determined over historical time. There were many gaps in this information and, because breeding ranges are in a constant state of flux, accurate and precise present-day ranges were not known before the Atlas project.

In addition to its main goal of providing up-to-date distribution maps of all the breeding species, the Atlas project attained several incidental goals, all of which contribute to better management of West Virginia's natural resources. The project has provided an inventory of the rare and endangered birds of West Virginia and has discovered new nesting locations for these species. Fragile and unusual habitats supporting rare species have been identified. The *West Virginia Breeding Bird Atlas* provides baseline data against which future changes in range and status of breeding birds can be compared. The sound factual database that now exists will help environmental planners to make wise decisions regarding resource use in West Virginia, and it provides baseline data for environmental impact statements. Finally, the Atlas project has involved the birders of West Virginia in a cooperative effort that contributed not only to their edification and understanding of birdlife but also to their enjoyment.

West Virginia lies near the limits (either northern or southern) of the ranges of many bird species, and the detailed distributional data of the Atlas contribute to an understanding of the biology and ecology of many of these species. The West Virginia Atlas record for a Yellow-bellied Flycatcher nest was a first breeding record for the state, and the Atlas records of a nest and adults feeding young for the White-throated Sparrow established a new southernmost breeding record for the species.

History of the Project

The West Virginia Breeding Bird Atlas Project was organized in 1983 under an agreement between the Nongame Wildlife Program of the West Virginia Department of Natural Resources* (DNR) and the Brooks Bird Club of Wheeling. The West Virginia Breeding Bird Atlas Project was first proposed to the Brooks Bird Club by A. R. Buckelew, Jr., of Bethany College and G. A. Hall of West Virginia University in 1982. On November 7, 1982, the Brooks Bird Club board of directors approved a resolution adopting a state breeding bird Atlas project as a priority research effort. Buckelew and Hall were designated to codirect the project and were instructed to seek a grant from the West Virginia DNR Nongame Wildlife Program to support the Atlas effort. The Nongame Wildlife Program approved their proposal for funding of the project in July 1983. Under the terms of the agreement, the Nongame Wildlife Program provided funds and data processing services for the project from 1983 through the fieldwork years and the period of final compilation of the Atlas results and preparation of this book in 1992. Further support for the publication of this book was provided by a grant from the National Fish and Wildlife Foundation. The Brooks Bird Club provided some administrative services and many volunteers for the fieldwork.

Organization

Codirectors Buckelew and Hall divided between them the tasks of writing the Atlas handbook, designing field cards and summary sheets, editing the Atlas newsletter, maintaining volunteer address lists, and supervising the work of district coordinators.

To test the field cards and get some experience in Atlas techniques, members of the Brooks Bird Club carried out some fieldwork under the direction of the codirectors in 1983. They covered one block in Hancock County and seven in Pocahontas County. Douglas Kibbe, ornithological adviser to the Vermont Breeding Bird Atlas Project, assisted by Sandra Brown, covered seven more blocks in Canaan Valley that year. Kibbe and

*The name of the Department of Natural Resources was changed to the Division of Natural Resources in 1989.

Brown provided the Atlas organizers with valuable advice on atlasing techniques and organization.

The state was divided into eleven reporting districts. To facilitate management of the volunteer field workers, district coordinators for each district were appointed. District coordinators held training sessions in their districts. In 1984 the formal project got under way.

Survey Grid

The basic unit used by most atlas projects in the United States has been the 7.5-minute topographic map published by the U.S. Geological Survey. These maps (quadrangles) represent an area of approximately 148 square kilometers (58 square miles). The area of West Virginia is covered by 508 of these maps. After the quadrangles that fall mostly outside of West Virginia's borders were eliminated, the 454 remaining quadrangles were included on the grid. (See appendix B for the names and location of the quadrangles specified for the Atlas project.) To provide a finer resolution, each quadrangle was divided into six blocks, each covering an area of approximately 25 square kilometers (10 square miles).

It was evident at the start of the West Virginia Atlas project that sufficient manpower was not available to cover all of the approximately 2,700 blocks in West Virginia. Since complete coverage was not possible, the selection of one priority block from each quadrangle would ensure even coverage of the state. By a random choice, the block on the southeast corner of each topographic sheet (block 6) was selected as a *priority block*. In 17 quadrangles along West Virginia state borders, all or most of block 6 fell outside of the state; in such cases, another block was usually designated as the priority block.

In addition to the priority blocks, three other types of nonpriority blocks were covered. First, a number of special blocks were selected for coverage because they were known to contain habitats of great interest. Special blocks included blocks in Tomlinson Run State Park, Oglebay Park, McClintic Wildlife Management Area, Beech Fork State Park, Panther State Forest, Cranberry Glades Botanical Area, Gaudineer Knob and the Gaudineer virgin spruce forest, Canaan

Valley, and several additional blocks in the Monongahela National Forest. Second, many volunteers covered additional blocks near their homes or favorite vacation areas. Finally, Brooks Bird Club outings also included some special blocks as well as priority blocks. A total of 516 blocks were targeted by the study.

The criteria for completion of coverage varied with habitat. For example, one might expect to find more species of birds in a block containing a mixture of old fields, farm land, and secondary-growth forest than one would find in a block covered entirely by mature northern hardwoods. A block was considered complete with the recording of 75 percent of the potential species for the block. District coordinators, being most familiar with the birds of their districts, set the level of acceptable coverage. In blocks in which it was difficult to set an acceptable level of coverage, volunteers were asked to continue making trips to the block until no new species were added on successive trips.

Field Records

Volunteers used field cards (see fig. 1) to record information on breeding birds observed on their assigned blocks. At the end of each season, volunteers filled out a summary sheet (see fig. 2) to submit their data to the district coordinator. Field cards were also submitted at the end of the season, but most of them were returned to the volunteers after they were checked by coordinators. If the project were ever to be repeated, the codirectors would keep the field cards because valuable statistics were lost when the cards were returned. For example, the total number of hours volunteers spent in the field is unknown for the West Virginia Atlas because many field cards could not be recovered.

FIELD CARD

WEST VIRGINIA BREEDING BIRD ATLAS

QUADRANGLE

Quad Name	Quad No.
BLOCK (circle one)	YEAR
NW CW SW NE CE SE 1 2 3 4 5 6	

Observers Name:

Address:

Phone:

BREEDING CRITERIA & CODES

POssible
- O Species observed but not in breeding habitat
- X Species heard or seen in breeding habitat

PRobable
- A Agitated behavior or anxiety calls
- P Pair seen
- T Bird holding territory
- C Courtship or copulation
- N Visiting probable nest site
- B Nest building by wrens or woodpeckers

COnfirmed
- NB Nest building
- PE Physiological evidence of breeding
- DD Distraction display
- UN Used nest
- FL Recently fledged young
- FS Parent with fecal sac
- FY Parent with food for young
- ON Adult leaving/entering nest site
- NE Nest with eggs
- NY Nest with young
- ST Seven territorial males

SPECIES	1ST	PO	PR	CO
Wood-Pewee, Eastern				
Flycatcher, Olive-sided*				
Lark, Horned				
Swallow, Tree				
Bank				
Rough-winged				
Barn				
Cliff				
Martin, Purple				
Jay, Blue				
Raven, Common*				
Crow, American				
Fish*				
Chickadee, Black-capped				
Carolina				
Titmouse, Tufted				
Nuthatch, White-breasted				
Red-breasted				
Creeper, Brown				
Wren, House				
Winter*				
Bewick's*				
Carolina				
Marsh (Long-billed)*				
Sedge (Short-billed Marsh)*				
Mockingbird, Northern				
Catbird, Gray				
Thrasher, Brown				
Robin, American				
Thrush, Wood				
Hermit				
Swainson's				
Veery				
Bluebird, Eastern				
Gnatcatcher, Blue-gray				
Kinglet, Golden-crowned				

FIGURE 1 PORTION OF ATLAS FIELD CARD

The summary sheets were forwarded to the project codirectors for final checking before data processing by DNR personnel in Elkins. Summary sheets were kept as permanent records of the fieldwork. At the end of the project, the district coordinators and codirectors checked all computer records against the original summary sheets for errors.

A third form used during the project, the verification form, is discussed below, under the heading "Verification of Rarer Species and Species of Unknown Status."

Breeding Criteria and Codes

Atlas workers used a set of breeding codes (see table 1) to indicate the breeding evidence found for each species in a block. Volunteers entered codes on field cards and then transferred them to the summary sheets at the end of each breeding season.

The breeding categories represented by the codes are generally uniform from one state to another. West Virginia dropped one "confirmed" code—*ST* for seven singing males—after the 1985 field season and established a new code, *M*, at the "probable" level of breeding evidence for this category. This change in breeding category was made to bring the West Virginia categories into agreement with those used by other states.

Atlas Workers

The district coordinator was responsible for assigning blocks to the volunteer field workers and for receiving the data on the summary sheets. In most cases, coordinators made the decisions about the adequacy of coverage and also had the responsibility of checking the accuracy of records for the asterisked species and other unusual records (see "Verification of Rarer Species").

Volunteers from the Brooks Bird Club, local Audubon chapters, and nature clubs

West Virginia Breeding Bird Atlas Project — Summary Sheet

Quadrangle Name	Block (circle one)	Year
	1 2 3 4 5 6	19/ /

Name

Address

Phone

IMPORTANT .. PLEASE READ FIRST
1. Transfer data from the small field record card to this form.
2. Use pencil to fill out this form.
3. Use one sheet per block.
4. For single letter codes use right hand box; use both boxes for double letter codes.

SPECIES	#	CODE		SPECIES	#	CODE		SPECIES	#	CODE	
Grebe, Pied-billed*	006			Woodpecker, Hairy	393			Parula, Northern	648		
Heron, Great Blue*	194			Downy	394			Warbler, Yellow	652		
Green-backed	201			Kingbird, Eastern	444			Magnolia	657		
Black-crowned Night*	202			Flycatcher, Great Crested	452			Black-throated Blue	654		
Yellow-crowned Night*	203			Phoebe, Eastern	456			Yellow-rumped*	655		
Bittern, Least*	191			Flycatcher, Yellow-bellied*	463			Black-throated Green	667		
American*	190			Acadian	465			Cerulean	658		
Goose, Canada	172			Alder	466			Blackburnian	662		
Mallard	132			Willow	850			Chestnut-sided	659		
Duck, Am. Black	133			Least	467			Yellow-thr.	663		
Teal, Green-winged*	139			Wood-Pewee, Eastern	461			Pine	671		
Blue-winged*	140			Flycatcher, Olive-sided*	459			Prairie	673		
Duck, Wood	144			Lark, Horned	474			Ovenbird	674		
Merganser, Hooded*	131			Swallow, Tree	614			Waterthrush, Northern	675		
Vulture, Turkey	325			Bank	616			Louisiana	676		
Black*	326			Rough-winged	617			Warbler, Kentucky	677		
Goshawk, Northern*	334			Barn	613			Mourning	679		
Hawk, Sharp-shinned*	332			Cliff	612			Yellowthroat, Com.	681		
Cooper's*	333			Martin, Purple	611			Chat, Yellow-breasted	683		
Red-tailed	337			Jay, Blue	477			Warbler, Hooded	684		
Red-shouldered	339			Raven, Common*	486			Canada	686		
Broad-winged	343			Crow, American	488			Redstart, American	687		
Eagle, Bald*	352			Fish*	490			Sparrow, House	688		
Golden*	349			Chickadee, Black-capped	735			Bobolink	494		
Harrier, Northern*	331			Carolina	736			Meadowlark, E.	501		
Osprey*	364			Titmouse, Tufted	731			Blackbird, Red-wgd.	498		
Kestrel, American	360			Nuthatch, White-breasted	727			Oriole, Orchard	506		
Grouse, Ruffed	300			Red-breasted	728			Northern	507		
Bobwhite, Northern	289			Creeper, Brown	726			Grackle, Com.	511		
Pheasant, Ring-necked	309			Wren, House	721			Cowbird, Brown-headed	495		
Turkey, Wild	310			Winter*	722			Tanager, Scarlet	608		
Rail, King*	208			Bewick's*	719			Summer	610		
Virginia*	212			Carolina	718			Cardinal, Northern	593		
Sora*	214			Marsh*	725			Grosbeak, Rose-br.	595		
Moorhen, Common*	219			Sedge*	724			Blue	597		
Coot, American*	221			Mockingbird, Northern	703			Bunting, Indigo	598		
Killdeer	273			Catbird, Gray	704			Dickcissel	604		
Woodcock, American	228			Thrasher, Brown	705			Finch, Purple	517		
Snipe, Common*	230			Robin, American	761			House	519		
Sandpiper, Upland*	261			Thrush, Wood	755			Siskin, Pine*	533		
Spotted	263			Hermit	759			Goldfinch, Amer.	529		
Dove, Rock	313			Swainson's	758			Crossbill, Red*	521		
Mourning	316			Veery	756			Towhee, Rufous-sided	587		
Cuckoo, Black-bill	388			Bluebird, Eastern	766			Sparrow, Savannah	542		
Yellow-billed	387			Gnatcatcher, Blue-gray	751			Grasshopper*	546		
Barn-Owl, Com.*	365			Kinglet, Golden-crowned	748			Henslow's*	547		
Screech-Owl, E.	373			Waxwing, Cedar	619			Vesper	540		
Owl, Great Horned	375			Starling, European	493			Lark*	552		
Barred	368			Vireo, White-eyed	631			Bachman's*	575		
Long-eared*	366			Yellow-throated	628			Junco, Dark-eyed	567		
N. Saw-whet*	372			Solitary	629			Sparrow, Chipping	560		

FIGURE 2   PORTION OF ATLAS SUMMARY SHEET

TABLE 1 BREEDING CODES AND DEFINITIONS

"Possible"

O   Species observed in the block but not in breeding habitat. This code was primarily for birds for which there was no evidence of breeding in the block. Species outside of the safe dates with no further evidence of breeding were also recorded as O.

X   Singing male present in suitable habitat during its breeding season.

"Probable" (always a one-letter code)

A   Agitated behavior or anxiety calls from adults. Parent birds responding to threats with distress calls or by attacking intruders. This did not include response to "squeaking," "pishing," or tape playing.

P   Pair observed in suitable breeding habitat within safe dates.

T   Permanent territory presumed through defense of territory. This included, for example, chasing other birds or song from the same location on two occasions at least a week apart.

C   Courtship or copulation observed. This included displays and courtship feeding.

N   Visiting probable nest site. This code applied primarily to hole nesters. It was also used when an atlaser observed a bird visiting a site repeatedly but saw no further evidence of breeding.

B   Nest building by wrens or excavation by woodpeckers.

M   Seven or more singing males in suitable breeding habitat within safe dates.

"Confirmed" (always a two-letter code)

DD   Distraction display, including injury feigning.

NB   Nest building by all species except wrens and woodpeckers. Carrying of sticks is part of the courtship ritual of some species (code C).

PE   Physiological evidence of breeding (i.e., highly vascularized incubation patch or egg in oviduct) based on the bird in hand. This code is used primarily by banders.

UN   Used nest.

FL   Recently fledged young incapable of flight (altricial species) or downy young (precocial species) restricted to the natal area by dependence on adults or limited mobility. Young cowbirds confirmed both cowbird and host species.

FS   Adult bird seen carrying fecal sack.

FY   Adult carrying food for young.

ON   Occupied nest; adults entering or leaving nest in circumstances indicating occupied nest. Primarily for high nests or nest holes, whose contents cannot be seen.

NE   Nest with eggs. (Cowbird eggs confirmed both cowbird and host.) Also included eggs or eggshells found on the ground.

NY   Nest with young seen or heard.

covered the blocks during one or more breeding seasons. Some volunteers responded to articles in newspapers. Codirector Buckelew was responsible for overall administration of the volunteers and record-keeping. District coordinators maintained contact with volunteers in their districts.

A newsletter, "West Virginia Breeding Bird Atlas News," edited by codirector Buckelew, was established to motivate, educate, and inform the volunteers. The newsletter appeared in 10 issues between the summer of 1984 and summer 1989. It contained articles on proper application of breeding codes, solutions to common atlasing problems encountered by volunteers, yearly progress reports, up-to-date names and addresses of coordinators, and due dates for summary sheets.

The district coordinators and codirectors met annually from 1985 through 1988. The meetings were used to discuss problems, assess the progress made each year, introduce new coordinators, and build enthusiasm for the project.

Timing of the Atlas Survey

Ideally, each species should be surveyed at the peak of its breeding season. For the majority of species nesting in West Virginia, this peak is from late May to mid-July, and most of the Atlas work was therefore done during this period. However, some species, notably owls, begin nesting in December or January, and other species, waterfowl and some permanent residents such as chickadees, are nesting in April. Special effort had to be made for these exceptional species. The most efficient time to make confirmations is in late June and early July; observing newly fledged young accompanied by their parents at this time is easier than finding nests earlier in the season. In the fall, after the leaves had fallen, some atlasers visited blocks to locate used nests.

A problem that arises both early and late in the breeding season is the differentiation between breeding individuals and migrants on their way to or from more northerly breeding grounds. For each species, a set of Atlas "safe dates" was established. Safe dates are defined as the period of time in which an observer could assume that a species seen in appropriate habitat was indeed a breeding bird and not a migrant. The safe dates obviously do not apply to "confirmed"

breeding records, which could be made outside of safe dates.

Atlas Handbook

Each Atlas volunteer was supplied with a copy of the "West Virginia Breeding Bird Atlas Project Handbook." Besides outlining the general instructions for the Atlas work and defining the breeding criteria and codes, the handbook contained a list of species expected to occur in the state. It also gave a brief statement of the nesting habitat, suspected distribution, and safe dates for each species.

Block Busting

Due to the uneven distribution of volunteers in the state, much of the state was covered by block busting. In block busting, an individual or a small group of highly skilled birders makes a special trip, spending a day or two in an intensive survey of a block or several blocks. Most block-busting trips took place in June or early July.

Verification of Rarer Species and Species of Unknown Status

Rare, threatened, or endangered species; colonial nesting waterbirds; and species whose status was in doubt were selected for more detailed reporting. These species were indicated on the summary sheets and field cards by an asterisk and were designated asterisked species. Fifty-four species, about 30 percent of West Virginia's 182 potential breeding species, were selected for this special reporting. These species are listed in table 2. After completion of the project it was clear that several additional species, such as the American Black Duck, Spotted Sandpiper, Alder Flycatcher, Warbling Vireo, and Dickcissel might have been added to this list. It was also evident that several of our asterisked species need not have been included.

Verification forms were required for all records of asterisked species. The forms requested information on the exact location, habitat description, details of breeding behavior observed, circumstances of the observation, details of identification, and the observer's previous experience with the bird.

Volunteers were encouraged to report asterisked species in any location, not only in priority blocks. The district coordinators were asked to verify all reports of asterisked species by using the verification form; the final acceptance of these reports was determined by the district coordinators.

Data Processing and Software

Data processing was performed by West Virginia Department of Natural Resources personnel at the Elkins Operations Center. The data are permanently stored on tape at the West Virginia Network for Educational Telecomputing (WVNET). Data were analyzed using the Statistical Analysis System (SAS Institute, Inc.). The species accounts maps were produced using Atlas Graphics (Strategic Mapping, Inc.), saved in Hewlett-Packard Graphics Language format, and converted to Encapsulated PostScript using HiJaak (Inset Systems, Inc.).

USFWS Breeding Bird Survey Data

The U.S. Fish and Wildlife Service (USFWS) Breeding Bird Survey (BBS) is a continent-wide project to monitor bird populations. The full details of the organization of the project and the methodology are given by Robbins, Bystrak, and Geissler (1986). Routes are randomly selected in each one-degree block of latitude and longitude. In running a route, a volunteer observer drives 24.5 miles, stopping every half mile for a three-minute observation period. During that time, all birds seen within a quarter mile and all birds heard are recorded. The routes are run during June and start 30 minutes before sunrise. These data are submitted to the Office of Migratory Bird Management of the U.S. Fish and Wildlife Service for analysis. In West Virginia, up to 40 routes are

TABLE 2   ASTERISKED SPECIES

Pied-billed Grebe	Chuck-will's-widow
American Bittern	Red-headed Woodpecker
Least Bittern	Yellow-bellied Sapsucker
Great Blue Heron	Olive-sided Flycatcher
Black-crowned Night-Heron	Yellow-bellied Flycatcher
Yellow-crowned Night-Heron	Fish Crow
Green-winged Teal	Common Raven
Blue-winged Teal	Bewick's Wren
Hooded Merganser	Winter Wren
Black Vulture	Sedge Wren
Osprey	Marsh Wren
Bald Eagle	Loggerhead Shrike
Northern Harrier	Golden-winged Warbler
Sharp-shinned Hawk	"Brewster's" Warbler
Cooper's Hawk	"Lawrence's" Warbler
Northern Goshawk	Nashville Warbler
Golden Eagle	Yellow-rumped Warbler
King Rail	Prothonotary Warbler
Virginia Rail	Swainson's Warbler
Sora	Bachman's Sparrow
Common Moorhen	Lark Sparrow
American Coot	Grasshopper Sparrow
Upland Sandpiper	Henslow's Sparrow
Common Snipe	Swamp Sparrow
Barn Owl	White-throated Sparrow
Long-eared Owl	Red Crossbill
Northern Saw-whet Owl	Pine Siskin

run each year, and the BBS project has been continuous since 1966.

The Atlas species accounts that make up the major portion of this book include BBS data for West Virginia, calculated as the annual rate of change of population (trend) by methods described in Robbins, Bystrak, and Geissler (1986). These trends, together with their statistical significance, were obtained from data from 1966 to 1989 as distributed to the state's USFWS Breeding Bird Survey coordinator by the Office of Migratory Bird Management.

Unpublished Atlas Data from Neighboring States

The states of Ohio, Pennsylvania, Maryland, and Virginia provided the West Virginia Breeding Bird Atlas Project with data from their own atlas projects that were in various stages of completion. Some of the data were used in the preparation of species accounts in this book, and this use of data is cited as "atlas data" in the species accounts.

Biogeographical Regions

Hall (1983) divided the state into three avifaunal regions: the Ridge and Valley Region, the Allegheny Mountains Region, and the Western Hills Region (see fig. 3). These three regions, which roughly correspond to both the physiographic and phytogeographic regions, have been adopted for the description of breeding ranges in this *Atlas*. Figure 4 shows a schematic representation of the topography of the state, as well as the principal river systems.

FIGURE 3   BIOGEOGRAPHICAL REGIONS OF WEST VIRGINIA

RIDGE AND VALLEY REGION

ALLEGHENY MOUNTAINS REGION

WESTERN HILL REGION

Ridge and Valley Region

The Ridge and Valley Region includes the easternmost counties of the state from the Virginia line on the east, west to the sharp escarpment formed by the Allegheny Front and Allegheny Mountain from Mineral County south through Virginia to Greenbrier County. A small portion of Monroe and Mercer counties is also in this region. As its name implies, the area is essentially a lowland, above which rises a series of long ridges, sometimes crowded and sometimes widely spaced. Both the ridges and the valleys have a northeast to southwest orientation (see fig. 4). The elevations of the westernmost ridges approach 1,800 meters, but the eastern ridges are generally lower, declining to about 700 meters. The valleys usually are about 400 meters lower, and the lowest elevation in the state (75.3 m)

is reached where the Potomac River leaves the state in Jefferson County.

The climax cover was largely oak-hickory-pine forest, although American chestnut was originally an important component. The forest in the easternmost portion of the Ridge and Valley Region was Appalachian oak type (see fig. 5). The valleys are largely unwooded, and agriculture is extensive, particularly in the broad Shenandoah Valley, a portion of the Appalachian Great Valley.

Allegheny Mountains Region

The region bounded by the Allegheny Front and Back Allegheny Mountain on the east and the line of Chestnut Ridge and Laurel Ridge south to the junction of Cheat Mountain and Back Allegheny Mountain in Pocahontas County is known as the Alle-

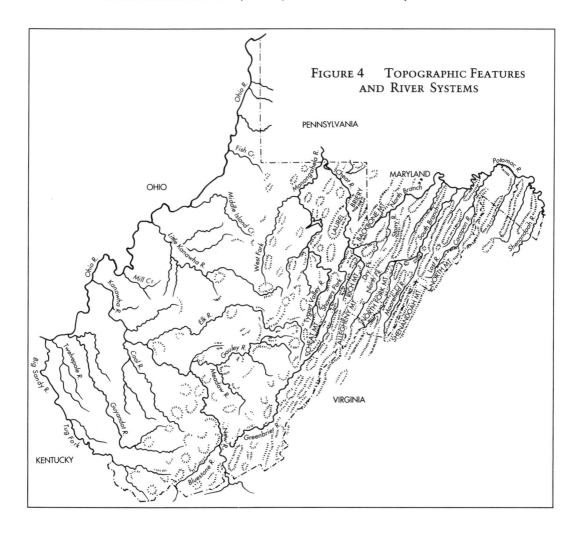

FIGURE 4   TOPOGRAPHIC FEATURES
AND RIVER SYSTEMS

gheny Mountains. The highest point in the state, Spruce Knob on Spruce Mountain in Pendleton County, reaches 1,482 meters, and many other ridges and knobs reach elevations above 1,300 meters. This region largely encompasses the headwaters of the Monongahela River system and, on its eastern slopes, the headwaters of the Potomac system.

The higher elevations were originally covered by a subalpine forest that in this region was an almost pure stand of red spruce, only 10 percent of which remains today. At the lower elevations, the forest is of the northern hardwoods type, with a band of mixed forest between the major formations (see fig. 5). The valleys of the Allegheny Mountains Region tend to be narrow and steep-sided, with agriculture marginally developed.

Western Hills Region

The remaining, and the largest, part of the state lies on the Unglaciated Allegheny Plateau, here called the Western Hills Region. The terrain is highly dissected in a dendritic pattern, and the region is characterized by steep hills and narrow valleys. Most of the hills are below 450 meters in height, and maximum relief usually ranges from 150 to 250 meters. In the southern portion of the region, the hills are higher and elevations occasionally reach 1,000 meters, with a few isolated knobs to 1,070 meters.

Vegetation in the Western Hills Region is usually described as central hardwoods. This broad designation contains several communities ranging from dry to wet: oak-pine, oak-chestnut (xeric), cove hardwoods or mixed mesophytic (mesic), and flood plain

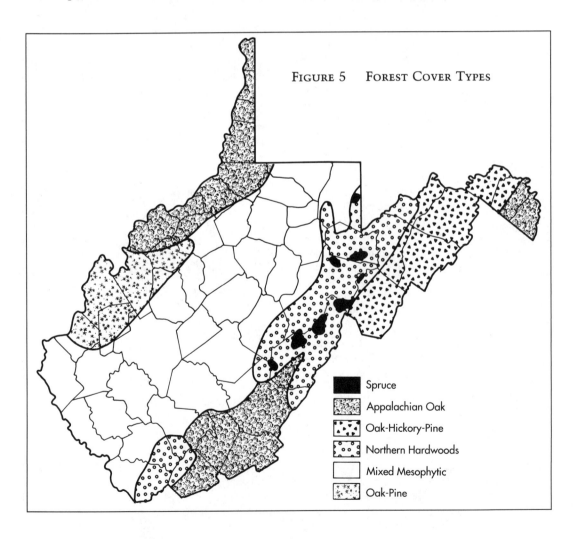

FIGURE 5   FOREST COVER TYPES

Spruce
Appalachian Oak
Oak-Hickory-Pine
Northern Hardwoods
Mixed Mesophytic
Oak-Pine

(hydric) (Strausbaugh and Core 1970–77). The northern hardwoods forest type occurs at the higher elevations in the southern hills in this region (see fig. 5).

Overview

The West Virginia landscape can be briefly described as hilly and wooded. Figure 6 shows an exaggerated east-west profile across the state. Recent survey data show that approximately 78 percent of the state is wooded (DiGiovanni 1990).

Figure 7 shows the percentage of forest cover by county. In the center of the state, more than 80 percent of the land area is forested. The age of the forest stands varies from a few remnant original stands, through mature, to early pole and sapling stages.

Geography and Climate

Variation in elevation and the resulting differences in moisture and temperature no doubt have an influence on bird distribution. No direct causal relationship should be drawn from these factors, however; the causes of bird habitat selection and distribution are likely to be complex.

The prevailing winds in West Virginia are westerly. The Allegheny Mountains have their ridges running from south-southwest to north-northeast in the eastern part of the state, and the forced ascent of warmer moisture-laden air and cooling air as it rises

from west to east causes the largest part of the state to the west of the mountain axis to receive greater amounts of precipitation. Immediately east of the mountains, a rain-shadow effect results in lower rainfall. At Pickens, Randolph County, high on the western slopes, the normal precipitation is 66 inches a year, whereas at Moorefield, Hardy County, it is only 31 inches (Core 1966). The mean annual temperature for the month of July is indicated in isolines in figure 8. Figure 9 summarizes normal annual precipitation classes by counties in West Virginia.

Population centers are indicated in figure 10, which also gives the names of counties. Distribution of birds, such as the Ovenbird, requiring large areas of wooded habitat may be negatively influenced by the presence of urban development. In contrast, the night-hawk and other birds that often nest in large towns and cities can be seen to be associated with urban areas.

Results and Discussion
Coverage

Thirteen of the 454 quadrangles in West Virginia had no blocks covered. In 21 quadrangles, the priority block was not block 6, since that block either fell partly or completely outside of the state or was inaccessible. In some of these 21 quadrangles, block 6 received some coverage anyway.

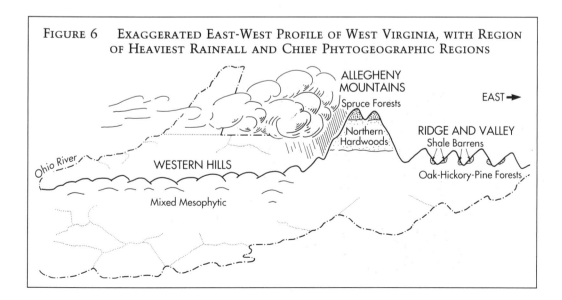

FIGURE 6   EXAGGERATED EAST-WEST PROFILE OF WEST VIRGINIA, WITH REGION OF HEAVIEST RAINFALL AND CHIEF PHYTOGEOGRAPHIC REGIONS

Introduction

Of the 424 number 6 blocks covered, 37 percent had more than 70 species listed, 49.5 percent had more than 50 species, and 12.5 percent had more than 20 species listed. There were 34,447 observations made. Of these, 34.6 percent were "confirmed," 30.6 percent were "probable," and 34.9 percent were "possible."

Coverage was not even across the state (see fig. 11). This is evident in spite of additional block-busting efforts and the addition of another year of fieldwork. (As 1988, the last proposed season, approached, project leaders realized that the survey could not be completed in the final year, and an additional year of fieldwork was added in 1989.) The portion of the state that received less coverage formed a corridor running from Mingo and McDowell counties northeast to Tyler, Doddridge, and Harrison counties; much of the central Allegheny Mountains

Region, comprising Webster, Upshur, and the western parts of Greenbrier, Pocahontas, and Randolph counties; and several priority blocks in Grant, Hardy, and Hampshire counties. Coverage of some of these areas was limited mainly to block-busting efforts because few Atlas volunteers lived in or near them. Some of the central portion of the state is sparsely populated, and in many cases, access to certain blocks was difficult, in part due to restrictions imposed by land owners. Some of the blocks in the urban setting near Charleston may have been ignored because of a lack of birds considered interesting by volunteers in the urban setting.

Limitations and Biases

The maps that accompany the species accounts in this *Atlas* represent our best current understanding about the breeding range

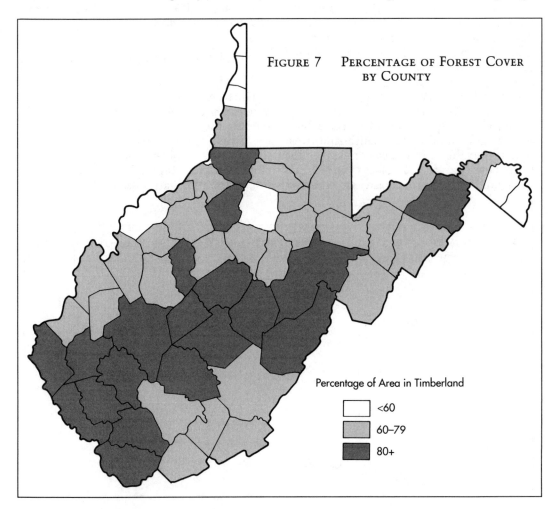

FIGURE 7   PERCENTAGE OF FOREST COVER BY COUNTY

Percentage of Area in Timberland

☐ <60

▨ 60–79

▪ 80+

of the bird species in West Virginia. However, users of this *Atlas* must recognize that there are limitations to the ranges as given. One cannot assume that a species not mapped in a given quadrangle or block is actually absent, since there are several reasons why a bird might not be recorded even though present.

Some breeding ranges are distorted because only one block in each quadrangle was surveyed. For species with a nearly statewide distribution, this is of little importance since the range is pretty well outlined on the map. But the breeding range of a species with limited distribution may appear to be smaller than it actually is; a species may have nested in other blocks within a given quadrangle but not in block 6. This could occur, for example, if the entire accessible area of block 6 were covered by forest. In such a case, no grassland birds would appear in that quadrangle, even though neighboring blocks may have had grassland that would support grassland birds.

Another reason breeding ranges may be distorted is that the amount of time spent on a block varied widely. Some were sampled over several years, whereas others were the subject of only a brief block-busting outing. The number of species recorded in a block increases, nonlinearly, with the amount of coverage. Although block busting often yielded long species lists because the volunteers participating in this procedure were usually highly skilled, a smaller percentage of species on these lists was "confirmed." In addition, nocturnal species and species that breed earlier in the season were often missed during block-busting trips, which frequently occurred only in early summer and during daylight hours.

Many blocks were covered in only one sea-

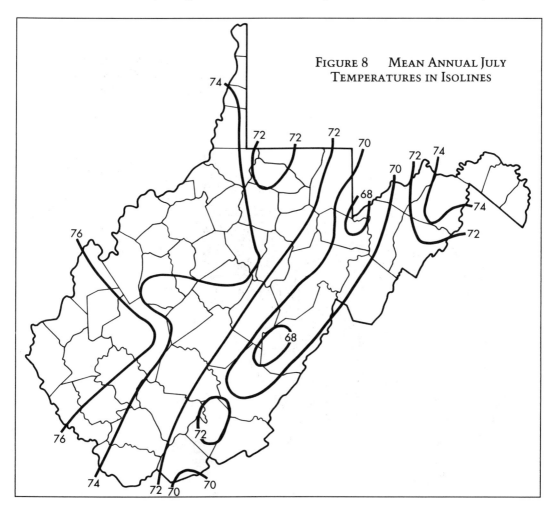

FIGURE 8   MEAN ANNUAL JULY
TEMPERATURES IN ISOLINES

son. Depending on the conditions that prevailed during that year, certain species may or may not have been found in the block. Thus, the Dickcissel appeared as a breeding bird in only one season, when a mass influx occurred from the western range, which was then undergoing a drought. Had the same Atlas blocks been surveyed only in some other year, the Dickcissel would have been missed. Similarly, Pine Siskins were found breeding only in summers following massive winter invasions. In other cases, species may have been missed because of specific conditions during a season. The White-throated Sparrow, for example, was present in a boggy area in one season but was not there the next year when the bog was dry.

Most of the blocks were surveyed during the months of June and July, and in the morning hours of daylight. Thus many early

nesting species, such as the Great Horned Owl and some of the raptors, may have been overlooked. Nocturnal birds, such as Whip-poor-wills and owls, may also have been missed. In an effort to improve the data for several nocturnal and easily identified species, advertisements were placed in local newspapers throughout the state soliciting information about Whip-poor-wills, Barn Owls, and Great Horned Owls. The return was moderately successful.

The ability of the volunteers to identify birds varied widely, and as a result some errors, both from misidentification and from failure to recognize the presence of a species, are in the data. Every effort was made by the district coordinators and the project codirectors to check and verify the reports of unusual or unexpected species. A species that was unfamiliar to a volunteer in a given

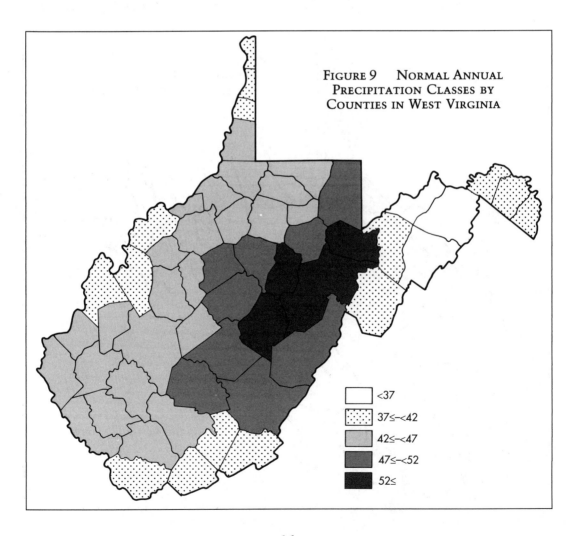

FIGURE 9   NORMAL ANNUAL PRECIPITATION CLASSES BY COUNTIES IN WEST VIRGINIA

<37

37≤–<42

42≤–<47

47≤–<52

52≤

block may not have been detected, and some volunteers may not have been able to hear certain high-pitched songs.

Weather

Weather conditions varied widely during the six years of the Atlas fieldwork. The wet and cool spring of 1984 was followed by a dry, warm June and a wet, cool July. Spring of 1985 was warm and wetter than normal, but the summer was cool with normal precipitation. Spring of 1986 was warmer than normal, and dry weather persisted into the summer. Temperatures were normal in June of that year, whereas July was warmer than normal. The following spring, 1987, which was warm and wet, was followed by a hot June and a dry July. The spring and summer of 1988 were abnormally dry and the summer was hot. A wet, cool spring in 1989 was followed by a warm July with normal precipitation.

Only the hot, dry season of 1988 had a serious impact on Atlas success. Many volunteers reported a lack of second nesting attempts in many species, and migrants were reported to have left their breeding territories earlier in the season than normal. Wetland species were so severely affected that some of them failed even to appear, and so they were missed in those blocks surveyed only in 1988. (For one such example, see the White-throated Sparrow account.)

Breeding Ranges

The ranges of most species as indicated by the Atlas data agreed with the ranges given in *West Virginia Birds* (Hall 1983). The new data did, however, fill in many of the gaps in

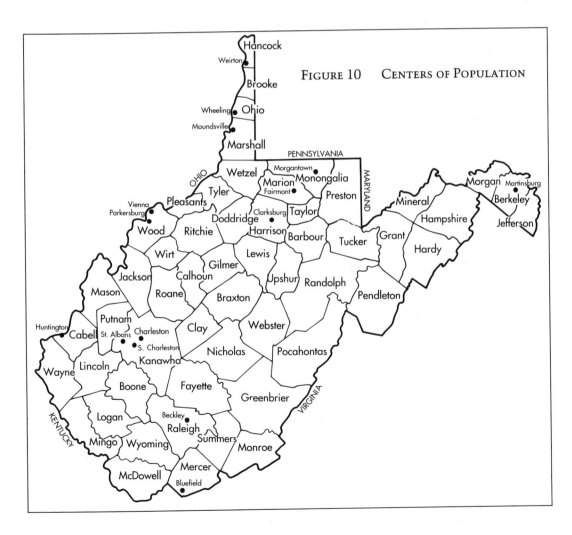

FIGURE 10   CENTERS OF POPULATION

Introduction

the knowledge available when that work was published. However, the extent to which northern and mountain species breed in the southern part of the Western Hills was not entirely expected. Species such as the Veery, Solitary Vireo, Black-throated Blue Warbler, and Dark-eyed Junco were found farther south and west than the previous data indicated.

The long-term trend of some northern species to expand their ranges southward continues, and during the Atlas period, the first breeding records in West Virginia for the Yellow-bellied Flycatcher and Yellow-rumped Warbler were established. A few southern species are also extending their ranges northward, as the Atlas results demonstrated for the Chuck-will's-widow and the Blue Grosbeak.

Widespread Species

The birds with the greatest distribution across West Virginia (see table 3) were species of edge and forest habitats, reflecting the dominant habitat types of the state. No grassland species are found in the top 15 most widespread species.

Unconfirmed Species

Atlas data failed to confirm 17 of the species on the project's list of potential breeders

TABLE 3 MOST WIDELY DISTRIBUTED SPECIES	
Species	No. of Blocks Reported
American Robin	515
Rufous-sided Towhee	515
Indigo Bunting	513
American Goldfinch	508
Song Sparrow	505
Red-eyed Vireo	502
Gray Catbird	499
American Crow	497
Chipping Sparrow	491
Wood Thrush	488
Tufted Titmouse	483
Scarlet Tanager	483
Eastern Phoebe	481
Northern Cardinal	480
Blue Jay	478

for the state (see table 4). Seven species which exhibited no probable evidence of breeding during the Atlas period were the Black-crowned Night-Heron, Yellow-crowned Night-Heron, Green-winged Teal, Golden Eagle, American Coot, Marsh Wren, and Red Crossbill. All seven are discussed

TABLE 4 SPECIES NOT CONFIRMED			
Species	PROBABLE	POSSIBLE	OBSERVED
American Bittern	1	5	2
Black-crowned Night-Heron	—	—	3
Yellow-crowned Night-Heron	(see Appendix A)		
Green-winged Teal	—	1	—
Northern Harrier	2	3	8
Golden Eagle	—	1	1
Common Moorhen	1	—	2
American Coot	—	1	—
Upland Sandpiper	1	—	—
Common Snipe	1	4	—
Chuck-will's-widow	3	4	—
Yellow-bellied Sapsucker	2	4	1
Olive-sided Flycatcher	3	—	—
Marsh Wren	—	1	—
Bachman's Sparrow	1	—	—
Lark Sparrow	2	—	—
Red Crossbill	—	1	5

in appendix A. The Peregrine Falcon, not found during the Atlas period but recorded as breeding in the state since that time, is also included in appendix A.

Suggestions for Future Work

The species accounts that follow report that many widespread and normally common species have shown marked declines in recent years. The reasons for this decline are not immediately apparent in most of these cases. Habitat deterioration is certainly one factor, but it is not the only one. Many forest-dwelling species, particularly the forest-interior species, have decreased in number as forest tracts have been fragmented. Many of these species are the so-called Neotropical migrants, which face threats at both ends of their migration. How-

ever, many of the common grassland and brushland species (Field Sparrow, Grasshopper Sparrow, Vesper Sparrow) are also declining. High priority should be given to monitoring these species and to researching the best management procedures.

In a somewhat different category, several once moderately common species have decreased in numbers until they are almost absent from West Virginia. The Barn Owl, Bewick's Wren, Loggerhead Shrike, Bachman's Sparrow, Lark Sparrow, and Henslow's Sparrow all merit continued monitoring.

The small number of Atlas records for wetland species emphasizes the need for protection of West Virginia's dwindling wetland resources. Such species as Pied-billed Grebe, American Bittern, Least Bittern, Hooded Merganser, King Rail, Virginia Rail, Sora, Common Moorhen, American Coot, Spotted

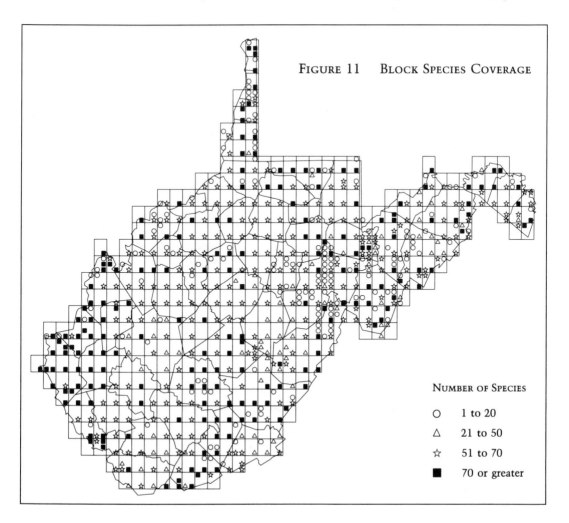

FIGURE 11   BLOCK SPECIES COVERAGE

NUMBER OF SPECIES

○   1 to 20
△   21 to 50
☆   51 to 70
■   70 or greater

Sandpiper, Common Snipe, and Sedge Wren should be considered for monitoring.

Other species that should be monitored include those at the periphery of their range in West Virginia, such as the Osprey, Bald Eagle, Northern Harrier, Northern Goshawk, Peregrine Falcon, Upland Sandpiper, Olive-sided Flycatcher, Yellow-bellied Flycatcher, Yellow-rumped Warbler, and Dickcissel.

Species Accounts

The species accounts that make up the main portion of *The West Virginia Breeding Bird Atlas* contain the following:

• A map showing the distribution of the species during the Atlas period.
• A brief summary of the species' continental breeding range, particularly in the northeastern portion of the range, often including a summary of the population trends as determined by the USFWS Breeding Bird Survey data (Robbins, Bystrak, and Geissler 1986).
• A summary of the species' West Virginia range, together with statements about the habitat occupied within that range.
• For most species, some general statements about nesting habits.
• Where appropriate, some comment on factors that affected the coverage.
• For most species, a population trend for the state, as determined by the BBS data.

Each account also includes, on the page with the distributional map, a table showing the number of blocks in which the species was recorded, together with the numbers of "confirmed," "probable," and "possible" breeding blocks.

Species Maps and Symbols

Each species account is accompanied by a map showing the distribution of the species as observed during the Atlas project. The species maps also show county outlines and a grid of the state's topographic maps. County names are given in figure 10, the state map of counties and population centers. The names of the topographic maps can be determined from the topographic map grid of West Virginia and the Key to Topographic Maps in appendix B.

Map symbols representing the status of the species records are located in the blocks surveyed. A circle represents an "observed" (O) record, a triangle a "possible" (X) record, a star a "probable" record, and a black square represents a "confirmed" record. (See the sample map in fig. 12).

State Maps

Plastic overlays of the state maps (figs. 3–5, 7–10) are provided in a slip envelope on the inside back cover, showing counties and the larger cities, mean annual temperature isotherms, normal precipitation by counties, percentage of area in timberland by county, biogeographical regions, forest types, and topographical features. The transparent overlay maps can be placed over species maps to study the occurrence of species related to features of West Virginia's geography and climate.

The West Virginia Breeding Bird Atlas offers a view of the birds breeding in the state during a six-year period in the late 1980s. It is not intended to function as a state bird book. Readers who would like more detailed information on the birds in this *Atlas* will find that *West Virginia Birds* (Hall 1983) provides a detailed account of the ecology of the state's birds. Included in this book are a history of West Virginia ornithology and a discussion of state avifaunal regions and habitat types. Hall's species accounts review nomenclature; abundance; residence status; migration dates; definite nesting records; location of specimens for rare or unusual species; quantitative data from breeding bird surveys, singing male censuses, and Christmas bird counts; and discussion of subspecies.

More information on nesting habits may be obtained from *A Field Guide to Birds' Nests* (Harrison 1975), and additional information on the ecology of North American birds may be found in *The Birder's Handbook* (Ehrlich et al. 1988).

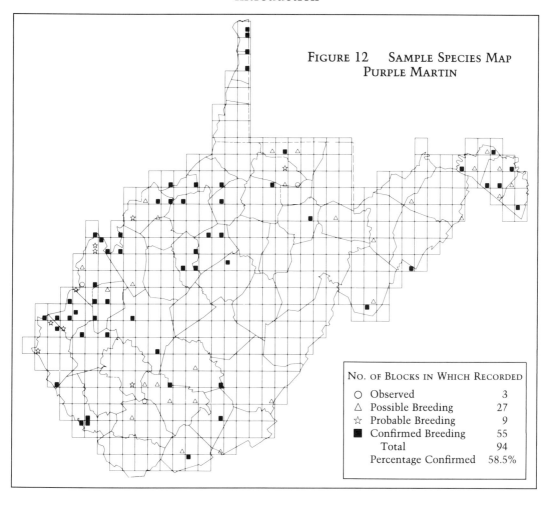

FIGURE 12   SAMPLE SPECIES MAP
PURPLE MARTIN

NO. OF BLOCKS IN WHICH RECORDED	
○ Observed	3
△ Possible Breeding	27
☆ Probable Breeding	9
■ Confirmed Breeding	55
Total	94
Percentage Confirmed	58.5%

Species Accounts

≪ ≫

Pied-billed Grebe *Podilymbus podiceps*

The PIED-BILLED GREBE inhabits marshy areas along streams and heavy emergent vegetation adjacent to ponds and lakes where there is sufficient open water for it to take flight. It breeds across all of the United States and southern Canada (AOU 1983). Formerly a fairly common but regular breeder in the Northeast, it may be declining in the region (Connor 1988). Hall (1983) noted several breeding records in West Virginia from the 1930s in Jefferson County, Ohio County, and Canaan Valley, and adults and young were present at McClintic Wildlife Management Area in 1972.

Atlas data show that the Pied-billed Grebe breeds in scattered locations in West Virginia. The majority of records are in the Ohio Valley. Birds were present at McClintic and Green Bottom wildlife management areas.

Atlas volunteers found fledged young near Hamlin in Lincoln County in three consecutive summers (1986–88).

After an early spring courtship, when the Pied-billed Grebe's loud, cuckoolike calls can be heard, the species is very secretive and hard to observe. Most Atlas workers took to the field later in the year, and some breeding may therefore have been missed.

This species may be as numerous in the state now as it has been at any time in this century (Hall 1983), but it should be included on the West Virginia list of vertebrate species of concern because of its noted decline in the Northeast, the scarcity of its wetland habitat, and its rarity in the state. Management techniques could include establishment of areas of emergent vegetation in recreational lakes and large ponds.

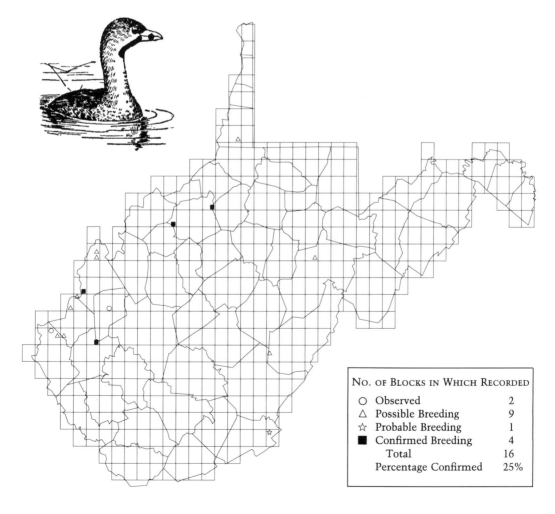

NO. OF BLOCKS IN WHICH RECORDED	
O Observed	2
△ Possible Breeding	9
☆ Probable Breeding	1
■ Confirmed Breeding	4
Total	16
Percentage Confirmed	25%

American Bittern *Botaurus lentiginosus*

The AMERICAN BITTERN inhabits secluded marshes, where it is most active at night and not easy to observe. Stealthy and quiet, it moves slowly through dense cattails, sedges, or rushes and at the slightest disturbance freezes in its characteristic upright, reedlike stance. This bittern's nest is usually placed on a mound above water or mud among tall emergent plants.

The American Bittern's breeding range includes most of North America, but a continentwide decline has occurred in recent years (Tate 1986). Hall (1983) described its occurrence as a rare to uncommon summer resident of marshy areas in West Virginia, noting regular summer records at Ashton and McClintic Wildlife Management Area in Mason County, Boaz Swamp in Wood County, and Altona Marsh in Jefferson County. He described it as quite numerous in Tucker County's Canaan Valley.

Atlas workers could not "confirm" breeding of this species in any of the areas mentioned by Hall. Some bitterns were probably missed because of their nocturnal, secretive habits and scarce habitat.

The American Bittern is listed as a species of concern by the West Virginia Department of Natural Resources (W. Va. DNR n.d.). A special effort should be made to census this species in Canaan Valley because this large wetland area is where most of the Atlas observations of bitterns were made.

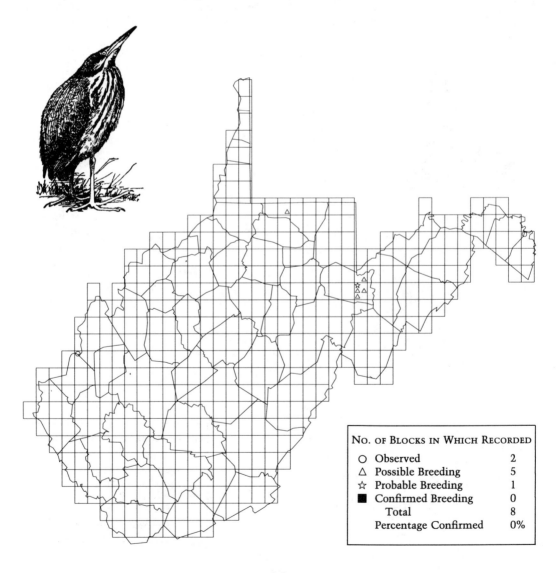

NO. OF BLOCKS IN WHICH RECORDED	
O Observed	2
△ Possible Breeding	5
☆ Probable Breeding	1
■ Confirmed Breeding	0
Total	8
Percentage Confirmed	0%

Least Bittern *Ixobrychus exilis*

The small, retiring LEAST BITTERN is a denizen of thick cattail or reedy marshes. Bent (1926) said that these birds may even be found in small pieces of marsh in the midst of a city, but they are seldom seen or heard, even when they nest in close proximity to human habitation. Least Bittern nests typically are placed about a foot above the water in a dense stand of cattails or somewhat higher in woody vegetation.

The Least Bittern breeds throughout most of North America and southern Canada. According to Hall (1983), the species is an un-common and very local summer resident, nesting regularly at Boaz Swamp, McClintic Wildlife Management Area, and Altona Marsh. The species is included on the West Virginia list of species of concern (W. Va. DNR n.d.).

Although Atlas workers "confirmed" breeding at McClintic and found it in three other widely scattered places, this bittern's actual status is at best uncertain. The bird is very difficult to find, so it is possible that some were overlooked.

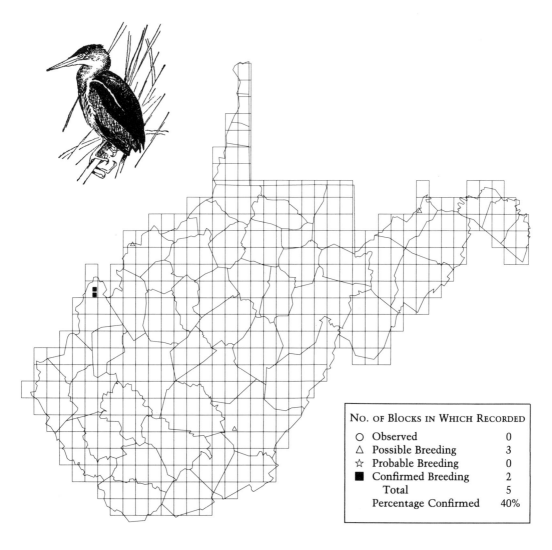

No. of Blocks in Which Recorded	
O Observed	0
△ Possible Breeding	3
☆ Probable Breeding	0
■ Confirmed Breeding	2
Total	5
Percentage Confirmed	40%

Great Blue Heron *Ardea herodias*

The GREAT BLUE HERON is widely distributed throughout the United States south of Canada. In West Virginia it is well distributed in the upper Ohio Valley, less common in the lower Ohio Valley (Kiff et al. 1986), and scarce in the Allegheny Mountains. It builds large platform nests near the tops of the largest trees, often on ridge tops. In other states, colonies of 100 or more nesting pairs are found, but heronries in West Virginia are smaller, usually with fewer than 20 nests. The colonies are conspicuous, especially in early spring before leaves appear. BBS trends for 1966–89 show an annual median 3.3 percent (p <0.05) increase in Great Blues reported in West Virginia.

Atlas workers found heronries at five locations along the Ohio River, Big Sandy River, and Tug Fork of the Big Sandy River. There are likely to be other colonies in neighboring Ohio, Pennsylvania, and Kentucky. The Atlas found very few of these herons in the interior of West Virginia; however, Rieffenberger (1988) discovered a colony of Great Blues on Cheat Mountain at 1,100 meters above sea level, and the cluster of Atlas records in Tucker County indicates the possibility of another colony in that area. There were scattered records in the Eastern Panhandle.

The Great Blue Heron often travels considerable distances between heronry and feeding grounds; distances as great as 16 kilometers have been recorded (Krebs 1974; Todd 1940). Scattered colonies could therefore account for many of the herons seen in the Ohio Valley during the Atlas period. Some of the scattered records in the interior and Eastern Panhandle may represent postbreeding and nonbreeding individuals.

Great Blue Herons are sensitive to human disturbance and avoid placing colonies near human settlement. Special efforts should be made to protect nest colonies. Great Blue Heron colonies are monitored in West Virginia by the DNR Nongame Wildlife Program. The recent establishment of the Ohio River Islands National Wildlife Refuge may provide some protection to colonies on refuge land.

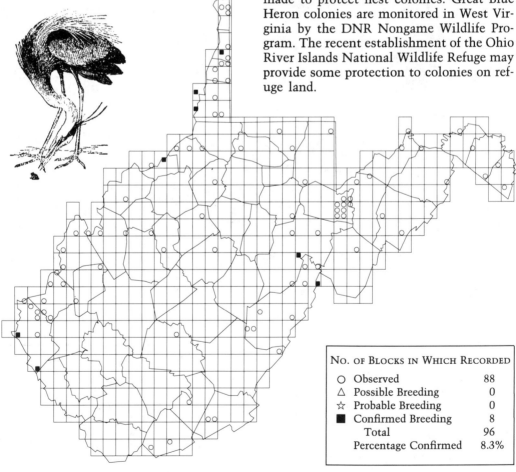

NO. OF BLOCKS IN WHICH RECORDED	
O Observed	88
△ Possible Breeding	0
☆ Probable Breeding	0
■ Confirmed Breeding	8
Total	96
Percentage Confirmed	8.3%

Green Heron *Butorides virescens*

The most common heron in West Virginia, the GREEN HERON frequents creeks and rivers at lower altitudes. It appears to be absent from tributaries of the Coal River in Boone County, tributaries of the Mud and Guyandotte rivers in Lincoln County, and the upper Kanawha River in Kanawha County. Lack of secluded habitat in the Charleston area could account for the heron's absence in the more heavily populated Kanawha County. The swift upland streams of Boone and Lincoln counties presumably do not provide good heron habitat. Pollution of streams in the coal fields of southern counties may also limit the distribution of the species in that region. This species is also scarce in the mountains to the east, but it does occur in the lower valleys of the New, Greenbrier, Tygart, Cheat, Potomac, and Shenandoah rivers.

The Green Heron nests singly, or occasionally in loose colonies, in brush near streams or in trees in nearby forest. Its nest is a platform of sticks placed at various heights in trees or thick brush. This species generally nests close to streams or wetlands, and it rarely strays far during the breeding season.

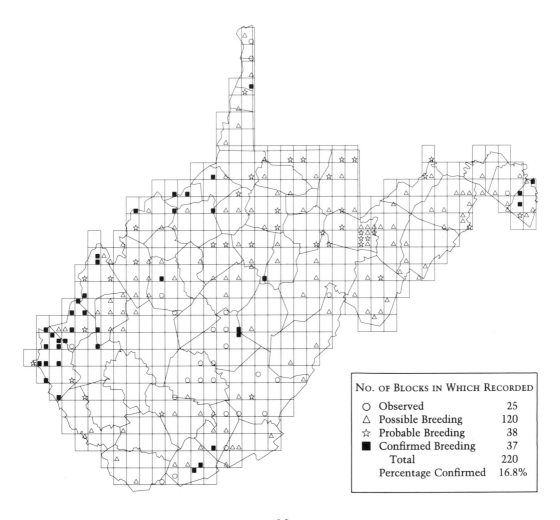

NO. OF BLOCKS IN WHICH RECORDED	
O Observed	25
△ Possible Breeding	120
☆ Probable Breeding	38
■ Confirmed Breeding	37
Total	220
Percentage Confirmed	16.8%

Canada Goose *Branta canadensis*

The CANADA GOOSE, having always been a migrant, is a relatively recent addition to the breeding avifauna of West Virginia. There are no records of the species breeding in the state prior to 1955, but since then it has been introduced by the Department of Natural Resources (DNR) as a game bird at various times in at least 22 counties (Hall 1983). It is now a regular breeder and permanent resident in many places. The heaviest concentrations appear to be in the major river valleys, the lower Ohio, the Tygart Valley, and the South Branch of the Potomac, where the major introductions were made by the DNR.

The Canada Goose is semidomesticated, and it tolerates human presence well as long as it has a nesting place safe from predators. The bird is often found in city parks and on farm ponds, and Atlas volunteers were able to "confirm" breeding for a high proportion of Canada Goose reports.

West Virginia Breeding Bird Survey workers recorded an average 9 percent (p <0.01) per year increase in Canada Geese counted for the years 1966 through 1989, and BBS data showed that Canada Geese increased at the rate of 9.5 percent (p <0.05) per year during the 1980s. The population of these birds is likely to continue to increase as long as suitable habitat remains for expansion. The Canada Goose is already considered a pest in some places, especially on golf courses and in parks.

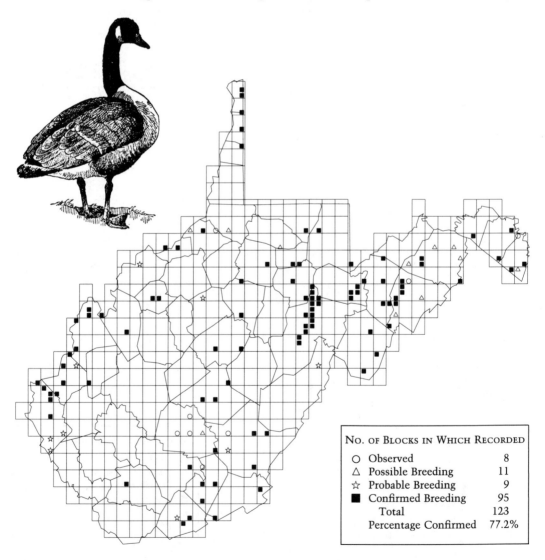

NO. OF BLOCKS IN WHICH RECORDED	
O Observed	8
△ Possible Breeding	11
☆ Probable Breeding	9
■ Confirmed Breeding	95
Total	123
Percentage Confirmed	77.2%

Wood Duck *Aix sponsa*

The WOOD DUCK breeds throughout the eastern United States east of the Great Plains. Its favorite nesting site is a fairly large natural cavity in a tree or an old Pileated Woodpecker hole. By the early 1900s, Wood Ducks were rare in the East, but conservation efforts, including hunting regulations and nest box programs, led to the gradual recovery of this species. Populations increased from the 1950s through the mid-1980s and then leveled off (Bellrose and Heister 1987).

In West Virginia the Wood Duck is the most widely distributed breeding duck, and it probably nests in every county. Hall (1983) lists 22 counties with nest records. Atlas workers found it in every county except Braxton, Clay, Gilmer, and Webster in cen-

tral West Virginia, Marshall County in the northwest, and McDowell County in the southern part of the state. The lack of Atlas records in the central counties is probably more of an indication of limited coverage than lack of habitat. For example, Brooks Bird Club Foray participants found several Wood Ducks, including a female with young, at the 1986 Glenville Foray in Gilmer County, but not in priority blocks in the same county (Ward 1987). The paucity of Atlas records in the mountain and southwestern counties is to be expected due to lack of habitat.

The future welfare of the Wood Duck depends on the maintenance of wooded bottomlands that provide the mast, wetland, and large nest trees it requires.

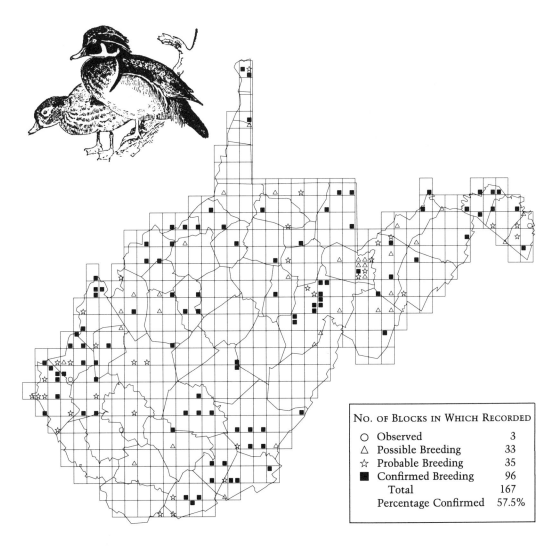

No. of Blocks in Which Recorded	
○ Observed	3
△ Possible Breeding	33
☆ Probable Breeding	35
■ Confirmed Breeding	96
Total	167
Percentage Confirmed	57.5%

American Black Duck *Anas rubripes*

Threatened by the continuing expansion of the Mallard into its northeastern North American range, the AMERICAN BLACK DUCK population has been decreasing since the 1950s. Hybridization with Mallards has gradually diluted the gene pool of the Black Duck. Some of the hybrids are difficult to separate from Mallards (Johnsgard 1975).

In West Virginia, the Black Duck summers on beaver ponds in Canaan Valley and on the Allegheny Front, in swamps in the lower Ohio Valley, in Hancock County in the Northern Panhandle, and in the Eastern Panhandle. Hall (1983) lists definite nesting records from six counties.

Atlas workers found breeding Black Ducks in the lower Ohio Valley and in Hancock, Monongalia, Tucker, and Pocahontas counties. No "confirmed" records were re-ported along the Shenandoah River and North Branch of the Potomac River where Hall had reported the species nesting. Likewise, they were not found in the marshes of Jefferson and Berkeley counties. Black Ducks may have been underreported both because of the remoteness of mountain beaver ponds and because some of these ducks (especially hybrids) may have been misidentified as Mallards. Atlas workers were not asked to report Black Duck x Mallard hybrids, but comparison of the range maps of the two species shows that Mallards are found wherever Black Ducks breed in the state.

There seems to be little that can be done about the decline of the American Black Duck in the face of the continued increase in Mallards.

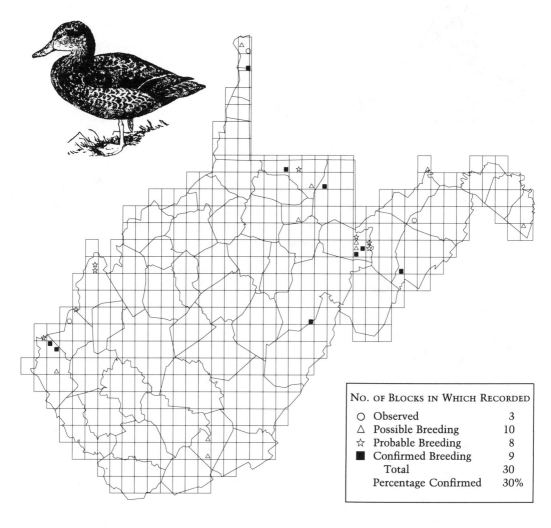

NO. OF BLOCKS IN WHICH RECORDED	
○ Observed	3
△ Possible Breeding	10
☆ Probable Breeding	8
■ Confirmed Breeding	9
Total	30
Percentage Confirmed	30%

Mallard *Anas platyrhynchos*

The familiar MALLARD breeds throughout northeastern North America, north to central Canada and northern New England, east to the middle Atlantic coast, and sparingly south of its main range. It is expanding northeast into American Black Duck range.

The Mallard places its nest in a depression on dry ground, sometimes far from water but usually near a small pond, lake, or stream. Hall (1983) reported that these ducks can be found anywhere in the state, but the Atlas project uncovered few records in the central Allegheny Mountains, the in-terior counties, and the southwestern hills. Mallards were found throughout the Ohio Valley, the New River Valley, the northern Allegheny Mountains, the Eastern Panhandle, and in various other scattered locations. Between 1966 and 1989, 64.3 percent ($p < 0.1$) of BBS routes in West Virginia reported increases in Mallard populations.

Mallards are often semidomesticated and are commonly found near human habitation. Thus, atlasers were able to report a high proportion (50.4%) of "confirmed" records.

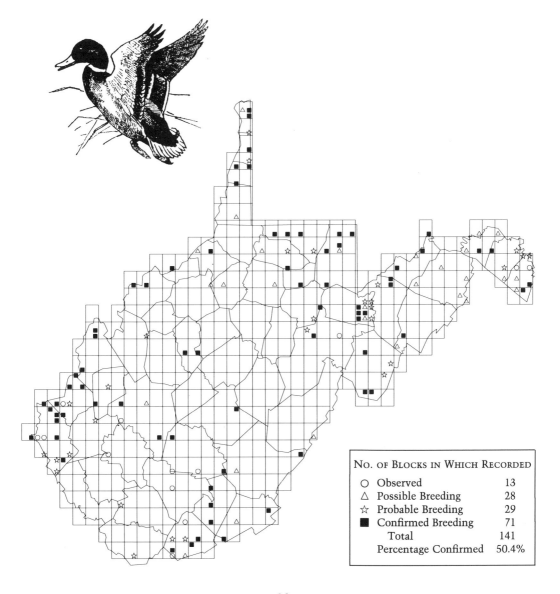

NO. OF BLOCKS IN WHICH RECORDED	
○ Observed	13
△ Possible Breeding	28
☆ Probable Breeding	29
■ Confirmed Breeding	71
Total	141
Percentage Confirmed	50.4%

Blue-winged Teal *Anas discors*

The attractive BLUE-WINGED TEAL builds its nest in a hollow in the ground, well concealed by surrounding vegetation. The nest is made of grasses or other plant material, mixed with down from the female's breast.

Teal breed across most of Canada and south casually to Tennessee and eastern North Carolina in the eastern United States (AOU 1983). Peterjohn (1989) noted a range expansion in Ohio ending in the 1970s, with fewer breeding pairs evident since 1980. This bird is a casual breeder in West Vir-

ginia, especially in the Ohio Valley (Hall 1983). In 1985 a pair nested at ponds at a shopping mall near Parkersburg (Hall 1985), and Atlas workers reported several pairs building nests on the shore of the lake at Beech Fork State Park in Wayne County; those nests were later destroyed by high water.

This species may breed more frequently in the lower Ohio Valley than was previously believed.

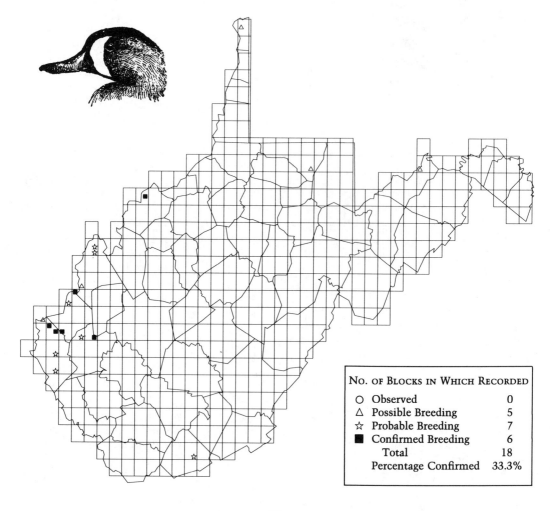

NO. OF BLOCKS IN WHICH RECORDED	
○ Observed	0
△ Possible Breeding	5
☆ Probable Breeding	7
■ Confirmed Breeding	6
Total	18
Percentage Confirmed	33.3%

Hooded Merganser *Lophodytes cucullatus*

The HOODED MERGANSER'S breeding range in the eastern United States extends from Florida through Georgia and the Appalachians, north to northern Canada, and westward to the central plains states. Hooded Mergansers prefer heavily wooded bottoms along swift, clear streams with nearby water less than 50 centimeters deep and close to timber. Although this species, like the Wood Duck, also uses tree cavities for nesting, competition for nest sites may not be terribly important because the Wood Duck prefers slow streams, backwaters, and ponds (Johnsgard 1975).

Hall (1983) noted breeding records at McClintic Wildlife Management Area, Mason County, Buffalo Creek in Brooke County, and on the Tygart Valley River near Elkins. The Atlas project found breeding Hooded Mergansers near Ashton in Mason County and on the Tygart near Valley Bend. There is apparently a small breeding population in the southern Ohio Valley and continued sporadic breeding in the upper Tygart Valley. Atlas workers may have missed some of these reclusive mergansers since their striking black-and-white pattern closely matches their surroundings in the deep shadows of heavily wooded bottoms.

Nest boxes placed in appropriate habitat might be effective in increasing numbers of this handsome duck.

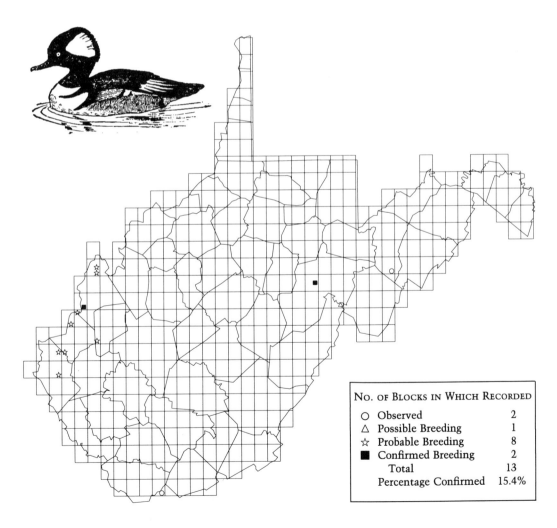

NO. OF BLOCKS IN WHICH RECORDED	
○ Observed	2
△ Possible Breeding	1
☆ Probable Breeding	8
■ Confirmed Breeding	2
Total	13
Percentage Confirmed	15.4%

Black Vulture *Coragyps atratus*

The breeding range of the BLACK VULTURE extends north from the Gulf Coast and Florida to central Ohio, southcentral Pennsylvania, and New Jersey (AOU 1983). In West Virginia it is an uncommon summer resident in the Eastern Panhandle and in the New River and Greenbrier River valleys (Hall 1983). Kiff et al. (1986) noted summer records in Wayne County, but they reported no evidence that the species remains in the lower Ohio Valley through the summer.

Atlas workers observed Black Vultures in 16 blocks in the Big Sandy River and Tug Fork valleys of Wayne and Mingo counties. However, vultures wander many miles from their breeding sites while foraging, so these scattered summer records could represent birds from nearby Ohio or Kentucky or nonbreeding birds. There may be a small breeding population in southeastern West Virginia, where the Atlas project found them in 23 blocks. Two nests with young were the only "confirmed" breeding records. One of these was discovered by Peregrine Falcon hacksite attendants in Grant County (Mitchell 1989); the other was found in Berkeley County.

Vulture nests are very difficult to find. Black Vulture nests are hidden in caves on cliffs, frequently in hollow stumps open only at the top, in hollow logs, or under fallen trees or brush. The birds sit very tight on the nest and are silent. Most nests are therefore found by accident.

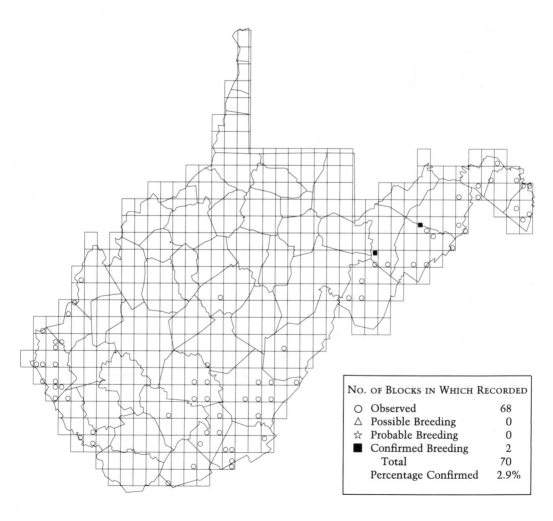

NO. OF BLOCKS IN WHICH RECORDED	
O Observed	68
△ Possible Breeding	0
☆ Probable Breeding	0
■ Confirmed Breeding	2
Total	70
Percentage Confirmed	2.9%

Turkey Vulture *Cathartes aura*

Formerly a more southern species, the TURKEY VULTURE has extended its range into southern Canada and New England during this century (AOU 1983). The reintroduction of the white-tailed deer and its subsequent overpopulation may be a factor in the range extension of this vulture (Sutton 1928), as are road kills and garbage dumps. A change in farming practices and the regrowth of forest in many areas may produce a varied habitat favorable to both deer and vultures. Turkey Vultures forage over miles of territory, primarily over open areas.

Turkey Vultures nest in situations similar to those of the Black Vulture and, similarly, most nests are found by accident. Turkey Vultures and Black Vultures often flock together, and their ranges overlap in West Virginia. Niche separation may be based on the Black Vulture's preference for larger prey and its lack of a sense of smell, which is keen in the Turkey Vulture (Stewart 1978).

After declining in the 1950s and 1960s, the Turkey Vulture population increased during the 1980s. The species is seen more frequently in the Morgantown area and the Northern Panhandle. BBS data show a 10.2 percent (p <0.01) per year increase since 1966 and a 18.1 percent (p <0.1) median annual increase from 1980 through 1989. Hall (1983) found the Turkey Vulture most common in southern and eastern West Virginia, but Atlas workers reported that it is now well distributed in the lower Ohio Valley and Northern Panhandle. It is less well distributed in the heavily wooded southcentral hill counties of McDowell, Wyoming, Logan, Boone, and Kanawha and in the northern counties of Wirt, Ritchie, Doddridge, Marion, and Harrison.

Many sight records are probably foraging breeders or nonbreeding birds, and it is not possible to assign them to a particular block for breeding. Only three were "confirmed," and they occurred in widely scattered places across the state.

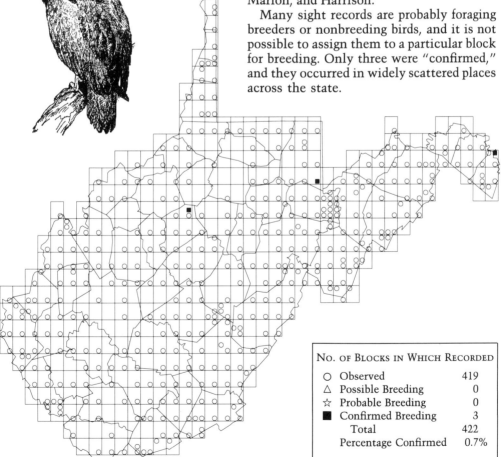

No. of Blocks in Which Recorded	
○ Observed	419
△ Possible Breeding	0
☆ Probable Breeding	0
■ Confirmed Breeding	3
Total	422
Percentage Confirmed	0.7%

Osprey *Pandion haliaetus*

The OSPREY was apparently never a common breeder in West Virginia. Eastern populations reach their western limit in West Virginia, and northern populations reach their southern limit in the Ohio River Valley. Hall (1983) documented six nest records from five counties. Populations of Osprey were greatly reduced by the effects of DDT during the 1950s and 1960s, but a period of recovery followed in the 1980s. Populations on the East Coast are approaching normal numbers, and the Osprey is no longer on the National Audubon Society's Blue List of species of concern (Tate and Tate 1982; Tate 1986).

The West Virginia Department of Natural Resources' Nongame Wildlife Program has sponsored a project to restore the Osprey to West Virginia by hacking young birds at three locations in the state. During the Atlas period, 41 young birds were released on the South Branch of the Potomac River in Hampshire County, 21 on Tygart Lake in Taylor County, and 8 on Blennerhasset Island in the Ohio River (Stihler pers. com. 1992). One of the pair that raised three young in Berkeley County in 1988 had been released in 1985 on the South Branch. Since the Atlas period, a nest was discovered on the West Virginia side of the North Branch of the Potomac River at Cumberland in 1990 (Stihler pers. com. 1992).

Some of the scattered summer Atlas records could be of released birds, and at least a few of the records from the Ohio Valley may be of migrating Ospreys. Migrants are fairly common in May and August along the Ohio River.

NO. OF BLOCKS IN WHICH RECORDED	
○ Observed	10
△ Possible Breeding	6
☆ Probable Breeding	1
■ Confirmed Breeding	1
Total	18
Percentage Confirmed	5.6%

Bald Eagle *Haliaeetus leucocephalus*

The BALD EAGLE nests most commonly along the shores of the Great Lakes and the eastern seacoast. Green (1985) reported that 110 pairs were breeding on the Chesapeake and lower reaches of the Potomac River approaching within 120 kilometers of West Virginia's Eastern Panhandle.

A nesting pair on the South Branch of the Potomac River, the first West Virginia breeding record in modern times, fledged two to three young each year from 1981 until the female was lost in 1988. A new nesting pair at an undisclosed site in Grant County fledged one young the same year (Hall 1988). Its location is not shown on our map for security reasons. Although there was no confirmation of young fledged at either nest in 1989 (Hall 1989), both nests produced young in 1990—two in the Grant County nest and three at the first nest location (Stihler pers. com. 1992). In 1991, three nests fledged five young (Hall 1991), and in 1992, four nests produced two eaglets each (Stihler pers. com. 1992).

Bald Eagle populations increased nationwide in the 1980s, and the species may be removed from the federal endangered species list in the near future.

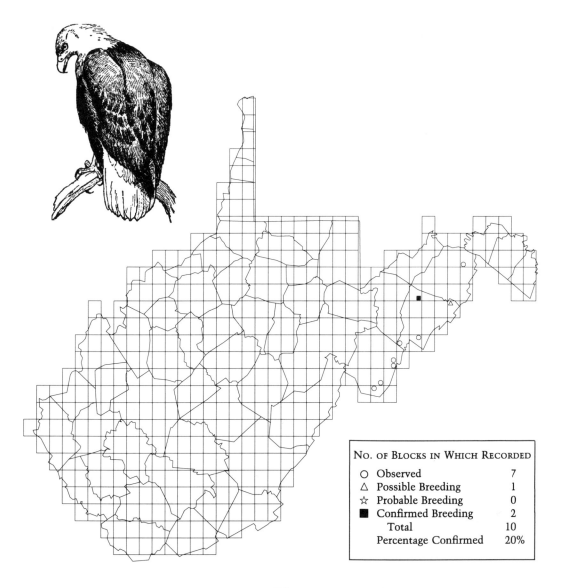

NO. OF BLOCKS IN WHICH RECORDED	
○ Observed	7
△ Possible Breeding	1
☆ Probable Breeding	0
■ Confirmed Breeding	2
Total	10
Percentage Confirmed	20%

Northern Harrier *Circus cyaneus*

Formerly called the Marsh Hawk, the NORTHERN HARRIER favors wet meadows and bogs, although it may nest occasionally in dry fields or reclaimed surface mines. Its nest is usually on the ground or on a mound of vegetation in a bog.

West Virginia is on the periphery of the harrier's range, which lies mainly to the north of the Ohio Valley and the Potomac River. It nests occasionally in mountain bogs and, prior to the 1950s, also nested at McClintic Wildlife Management Area (Hall 1983). This species has declined throughout this range since the 1950s (Tate 1986).

Atlas workers found a few summering harriers on the Allegheny Front. Several were also recorded in the lower Ohio River Valley, where they may continue to breed on occasion. No nesting of the species was discovered during the Atlas project.

Although the Northern Harrier is listed as a vulnerable species in Pennsylvania (Gill 1985), the Pennsylvania Atlas project discovered a number of new breeding localities for the species in the Allegheny Mountains, on reclaimed surface mines in Clarion County, and in the northcentral counties of that state (Brauning 1992). It is also listed in the *Vertebrate Species of Concern in West Virginia* (W. Va. DNR n.d.).

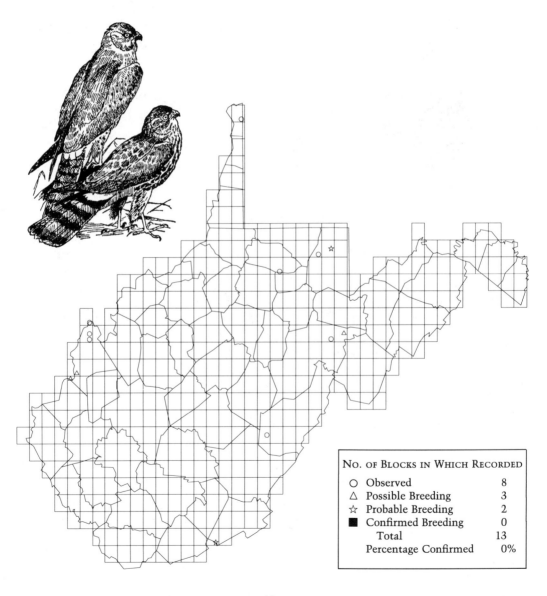

NO. OF BLOCKS IN WHICH RECORDED	
○ Observed	8
△ Possible Breeding	3
☆ Probable Breeding	2
■ Confirmed Breeding	0
Total	13
Percentage Confirmed	0%

Sharp-shinned Hawk *Accipiter striatus*

The SHARP-SHINNED HAWK breeds from central Ontario and Quebec, south to South Carolina (AOU 1983). According to Hall (1983), this hawk is probably most common in northern hardwoods forest at higher elevations in West Virginia.

As expected, volunteers found the Sharp-shinned Hawk in the Allegheny Mountains Region, and Atlas data also indicate it is more widespread than once thought in the southwestern counties. Atlas volunteers in eastern Kentucky also found more Sharp-shinned Hawks than had been expected there (Hall 1985). Very few were reported from the northwestern counties of West Virginia.

Sharp-shinned Hawks may be confused with the Cooper's Hawk, and the species may have been underreported because of the problems generally associated with finding forest-dwelling raptors. The Sharp-shinned Hawk's nest is often placed in thick conifers, where it may be hard to find.

This species declined precipitously in the early 1960s, probably due to egg-shell thinning from DDT. It remains on the National Audubon Society's Blue List of threatened species (Tate 1986), although increased numbers have been seen in migration in recent years. Observers in Pennsylvania have also reported a decline in the population of Sharp-shinned Hawks, which has been attributed to the loss of much of the coniferous woods in the state (Gill 1985). Likewise, U.S. Fish and Wildlife Service BBS trends for the years 1980 through 1989 show a 9 percent (p <0.01) annual decline in Sharp-shinned Hawk populations in West Virginia.

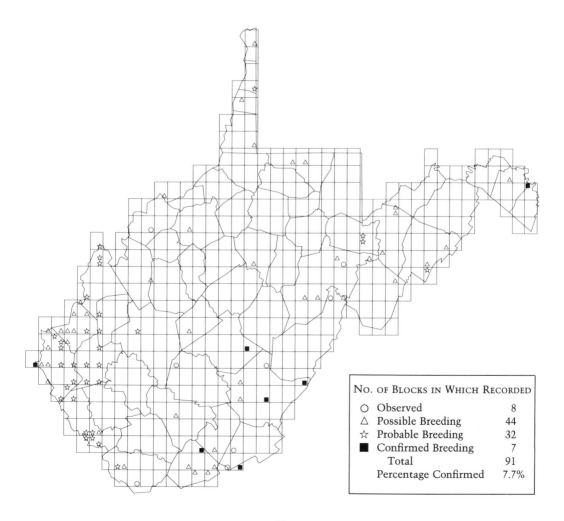

NO. OF BLOCKS IN WHICH RECORDED	
O Observed	8
△ Possible Breeding	44
☆ Probable Breeding	32
■ Confirmed Breeding	7
Total	91
Percentage Confirmed	7.7%

Cooper's Hawk *Accipiter cooperii*

Except for the southern tip of Florida, the COOPER'S HAWK is found throughout the United States north to southern Canada. In West Virginia, the species is less common in the mountains and is absent from spruce forest (Hall 1983). It is well distributed in the Northern Panhandle and southwestern counties. Atlas work has also shown the Cooper's Hawk to be more widespread in eastern Kentucky than was once thought (Hall 1989).

The Cooper's Hawk may not be reported as often as the Red-tailed Hawk and other soaring raptors. Absence of Atlas reports in the central counties may result from this difficulty or from a lack of coverage, but these factors cannot explain the species' scarcity in the Ohio River counties of Wood, Pleasants, and Tyler, or the far Eastern Panhandle, where coverage was generally good.

Formerly a very common raptor, the Cooper's Hawk has declined since the 1950s (Tate 1986). This hawk and other accipiters were adversely affected by the insecticide DDT. Accipiters may have been affected more than other raptors because their prey consists of birds, many of which eat insect prey (Snyder et al. 1973). The decline in the East seems to have ended, but this species remains on the National Audubon Society's Blue List of threatened species (Tate 1986), the List of Vulnerable Species in Pennsylvania (Gill 1985), and the *Vertebrate Species of Concern in West Virginia* (W. Va. DNR n.d.).

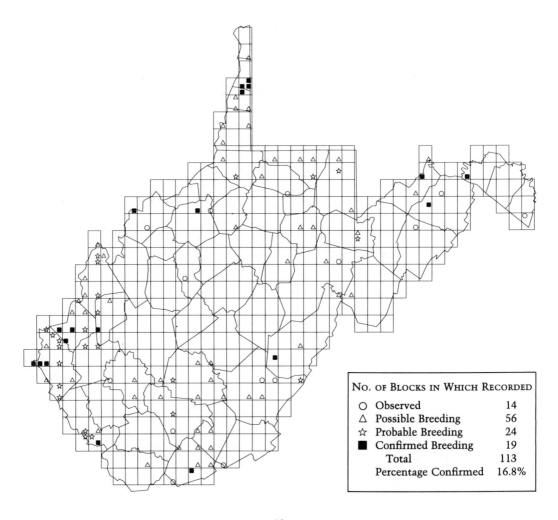

No. of Blocks in Which Recorded	
○ Observed	14
△ Possible Breeding	56
☆ Probable Breeding	24
■ Confirmed Breeding	19
Total	113
Percentage Confirmed	16.8%

Northern Goshawk *Accipiter gentilis*

Although in rare instances the NORTHERN GOSHAWK may breed as far south as the Great Smokies (AOU 1983), it generally reaches the southern extreme of its range in the West Virginia mountains, where it prefers hardwood forest habitat. Hall (1983) documented breeding records from Canaan Valley in Tucker County and Randolph County. The Atlas project extended the range southward by finding a nest with young in 1989 in northern Pocahontas County. Volunteers also recorded a summer sighting in the southern part of that county in 1987. In 1990, the year after the Atlas period ended, three Northern Goshawk records were noted: a nest with young northeast of Davis in Tucker County, a nest with young on Shaver's Fork in Pocahontas County, and two adults with two immature birds in Randolph County (Buckelew 1991).

The Goshawk places its nest of large sticks and twigs 9 to 12 meters high in a crotch or against the trunk of either deciduous or conifer trees. Northern Goshawks are noisy and conspicuous near the nest. They fearlessly dive and scream at intruders to drive them away. This behavior makes their nests relatively easy to find if an observer is near the nest.

The Northern Goshawk has been expanding its range in the Northeast since 1950, especially since the winter invasion of 1972 to 1973 (Gill 1985). The increase of forest cover in the past half-century may be a factor in this range expansion.

NO. OF BLOCKS IN WHICH RECORDED	
O Observed	0
△ Possible Breeding	1
☆ Probable Breeding	0
■ Confirmed Breeding	1
Total	2
Percentage Confirmed	50%

Red-shouldered Hawk *Buteo lineatus*

The RED-SHOULDERED HAWK requires large tracts of forest interspersed with small wetlands, where it obtains much of its prey. The nest is typically 9 to 18 meters from the ground in a main fork close to the trunk of a tall tree. The nest tree is often in the vicinity of water. Large open areas are excluded from Red-shouldered territories, and these hawks avoid buildings and roads, which are usually associated with more open habitat.

The Red-shouldered Hawk has declined throughout much of its range in the northeastern and Middle Atlantic United States (Tate and Tate 1982; Tate 1986). It is considered a vulnerable species in Pennsylvania (Gill 1985). Its decline is attributed to increased fragmentation of forest, drainage of wooded wetlands, and competition with the Red-tailed Hawk, which predominates in more open habitat (Bednarz and Dinsmore 1981).

Hall (1983) considered the Red-shouldered Hawk to be more common than the Red-tailed Hawk in most of West Virginia except in the Northern Panhandle. Atlas data show that, except in the northern Allegheny Mountains, the Red-shouldered Hawk is presently less widely distributed than is the Red-tailed Hawk.

To prevent further decline of this hawk, management of bottomland forest and other forest wetland habitat is needed. Mature forest containing small marshes, as well as backwater areas and forested wetland, should be maintained. Such management would also benefit the Wood Duck and other forest wetland species.

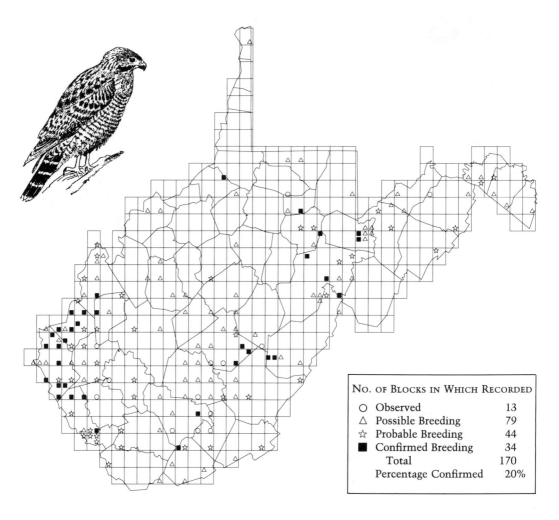

NO. OF BLOCKS IN WHICH RECORDED	
○ Observed	13
△ Possible Breeding	79
☆ Probable Breeding	44
■ Confirmed Breeding	34
Total	170
Percentage Confirmed	20%

Broad-winged Hawk *Buteo platypterus*

The BROAD-WINGED HAWK breeds in large tracts of continuous upland, deciduous forest throughout eastern North America. West Virginia Atlas workers found Broad-wings most often in the southern parts of the state and in the northern Allegheny Mountains Region. It was also well distributed in the Northern Panhandle, though volunteers reported it less often in the Ohio River Valley. The Ohio Atlas also shows few records in the Ohio Valley. The Broad-wing apparently prefers more upland habitat away from rivers.

Atlas findings did not support Hall's belief (1983) that the Broad-wing was the most common and widely distributed buteo in West Virginia. Whereas only 20 Broad-winged Hawks were "confirmed" in 258 blocks, the Atlas project "confirmed" 69 Red-tailed Hawks in 322 blocks where that species was present.

The Broad-winged Hawk is harder to observe than is the Red-tailed Hawk. Furthermore, the Broad-wing sits tight on its nest, which is often located in heavily wooded habitat, thus making it hard to "confirm." The nest is usually located 7 to 12 meters from the ground in the main crotch of a tree, often near small ponds or streams.

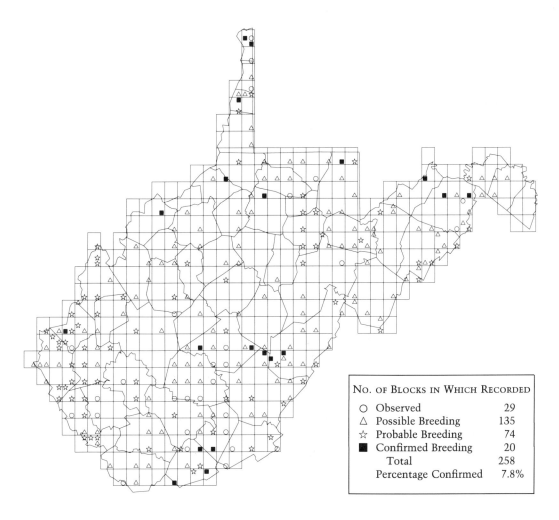

No. of Blocks in Which Recorded	
○ Observed	29
△ Possible Breeding	135
☆ Probable Breeding	74
■ Confirmed Breeding	20
Total	258
Percentage Confirmed	7.8%

Red-tailed Hawk *Buteo jamaicensis*

The RED-TAILED HAWK has become the most commonly encountered and most conspicuous buteo in the northeastern United States. Breeding bird surveys between 1965 and 1979 showed significant increases for this raptor in both western and eastern regions of North America (Robbins, Bystrak, and Geissler 1986). The Red-shouldered Hawk and Broad-winged Hawk are considered more common in most of West Virginia (Hall 1983), but Atlas workers found the Red-tailed Hawk in more blocks than either of these species. This is probably a reflection of the more heavily forested habitat of Red-shouldered and Broad-winged Hawks.

The Red-tailed Hawk places its large nest 10 to 27 meters from the ground in tall trees on the edges of woodlands. The nest is often reused and added to, becoming larger and more visible over the years. Its nest placement and its habit of soaring over open country makes the Red-tailed Hawk the most noticeable of West Virginia's large diurnal raptors. The extensive reporting of the species may be due in part to its conspicuous habits, but several other factors, such as forest fragmentation, may be contributing to the resurgence of this bird of more open habitats. When forests are cut, the more aggressive, larger Red-tailed Hawk usually replaces the Red-shouldered Hawk (Bednarz and Dinsmore 1981; Craighead and Craighead 1956). Furthermore, attitudes towards raptors have changed, and hawks are better protected today than they were in former years, when the Red-tailed Hawk was a frequent, easy target for the thoughtless gunner.

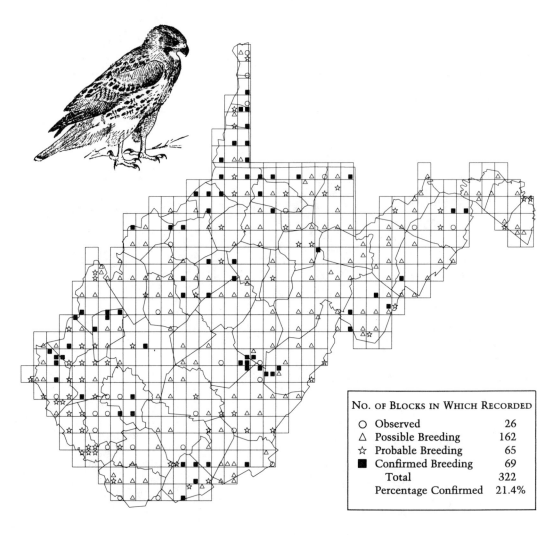

No. of Blocks in Which Recorded	
○ Observed	26
△ Possible Breeding	162
☆ Probable Breeding	65
■ Confirmed Breeding	69
Total	322
Percentage Confirmed	21.4%

American Kestrel *Falco sparverius*

This familiar raptor is widespread and common throughout North America. Its preference for open habitat limits its distribution in West Virginia, where Atlas workers found it throughout the Ohio Valley and in scattered locations in the north, in the Eastern Panhandle, and in the New River, Tygart Valley River, and Greenbrier River valleys. The kestrel is less common in the more heavily forested Allegheny Mountains and the southwestern Western Hills. The U.S. Fish and Wildlife Service BBS data showed a 4.3 percent ($p < 0.1$) increase per year in kestrel populations for the years 1966 through 1989, but in the last decade, there has been a 4 percent annual decrease in West Virginia.

Kestrels nest in natural cavities, flicker holes, nest boxes, and under eaves and in holes in buildings. Typically, nests are 3 to 10 meters from the ground. Kestrels readily become accustomed to the presence of people, and they are often seen perched conspicuously on wires in open country. Thus the species is relatively easy to observe. The Atlas map is probably a good estimate of the species' distribution in the state.

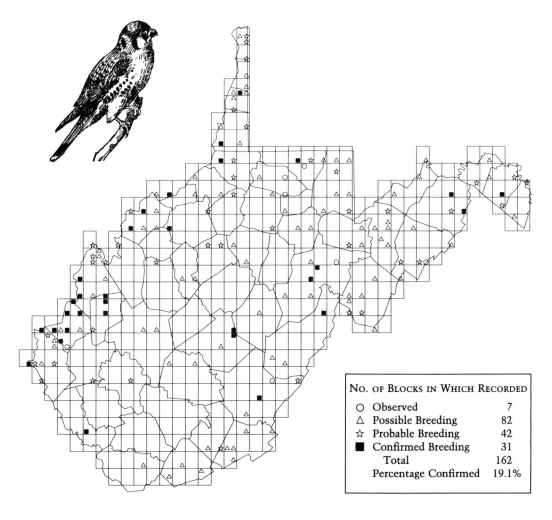

No. of Blocks in Which Recorded	
○ Observed	7
△ Possible Breeding	82
☆ Probable Breeding	42
■ Confirmed Breeding	31
Total	162
Percentage Confirmed	19.1%

Ring-necked Pheasant *Phasianus colchicus*

The RING-NECKED PHEASANT is a popular Asian game bird that has been widely introduced in the northeastern United States, where it nests in weedy pastures and hayfields. It reaches the southern limit of its established eastern range in the West Virginia Northern Panhandle, the far Eastern Panhandle, and other scattered locations in northern West Virginia. It has occasionally been introduced in other parts of the state without much success (Hall 1983).

Pheasants were more common in West Virginia prior to the severe winters of the 1970s. A general decline in pheasant populations since the 1940s is tied to changing agricultural practices, such as the elimination of hayfields and pastures, more frequent mowing of hayfields, fall plowing, and the abandonment of farming altogether in many places.

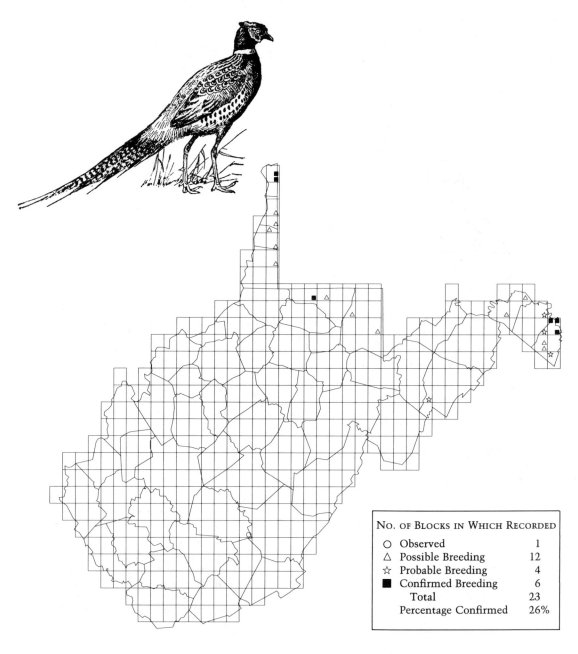

NO. OF BLOCKS IN WHICH RECORDED	
○ Observed	1
△ Possible Breeding	12
☆ Probable Breeding	4
■ Confirmed Breeding	6
Total	23
Percentage Confirmed	26%

Ruffed Grouse *Bonasa umbellus*

A popular game bird, the RUFFED GROUSE breeds in the Appalachians, eastern Ohio, New York, New Jersey, and north to Canada. It prefers second-growth deciduous or coniferous forest. Ruffed Grouse nest in thick woods and dense cover, where the female makes her nest in a hollow on the ground under a log or stump or in brush. Generally, the female flushes only when closely approached. Many "confirmed" breeding records were of fledged young.

Most forest in West Virginia has matured past the early successional second-growth forest favored by Ruffed Grouse. According to Crum (1989), the amount of prime habitat has declined precipitously, from about 22 percent in 1975 to 10 percent in 1989 of the total forest habitat available in the state.

Ruffed Grouse populations cycle through high and low numbers over periods of several years. West Virginia's Ruffed Grouse were at a high population level at the beginning of the Atlas period. In 1984, the first year of the Atlas project, populations peaked at about 1.8 flushes per hour statewide, and about 224 broods were reported to the Department of Natural Resources. Only 150 broods were reported in 1988, at the end of the Atlas period (Rieffenberger and Crum 1989). During the first three years of the Atlas period, when the grouse population was at its peak, the south-central and central parts of the state did not get much coverage. These areas might have shown better distribution of the species had they been covered during those years.

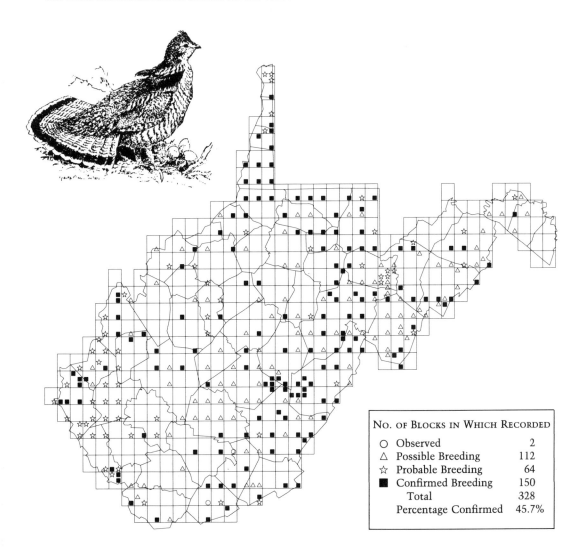

NO. OF BLOCKS IN WHICH RECORDED	
○ Observed	2
△ Possible Breeding	112
☆ Probable Breeding	64
■ Confirmed Breeding	150
Total	328
Percentage Confirmed	45.7%

Wild Turkey *Meleagris gallopavo*

Now restored to much of its former range, the WILD TURKEY is found from southern Ontario, Vermont, New Hampshire, and Maine, south to the Gulf Coast. As forests mature on abandoned farms, more turkey habitat becomes available. From 1961 to 1989, turkey habitat in West Virginia has increased from 50.2 percent to 63.7 percent of forest habitat (Crum 1989). The birds were hunted in all 55 counties in 1989, but the harvest was small in the southwestern counties (Pack 1989). This is reflected on the Atlas map, which indicates that no turkeys were observed over much of southwestern West Virginia. A Department of Natural Resources Wildlife Division turkey trap and transplant program had restored the species to 39 counties by 1989, and expansion of stockings made in the southwestern counties are expected to repopulate those counties as well. The statewide population of Wild Turkeys is estimated to exceed 70,000 (Pack 1989).

The Wild Turkey favors mature stands of deciduous forest where a plentiful supply of mast and berries exist. The birds also make forays into corn and grain fields. The female makes her nest in a depression on the ground, usually under dense brush at the base of a tree, under the top of a felled tree, or in a similar concealed place. The nests are hard to find and, consequently, most Atlas "confirmed" records were of fledged young. Of course, it is also true that Atlas volunteers were most likely to be in the field in May and June when females and chicks are present. Only nine Atlas records for the Wild Turkey were "confirmed" by finding the nest and eggs.

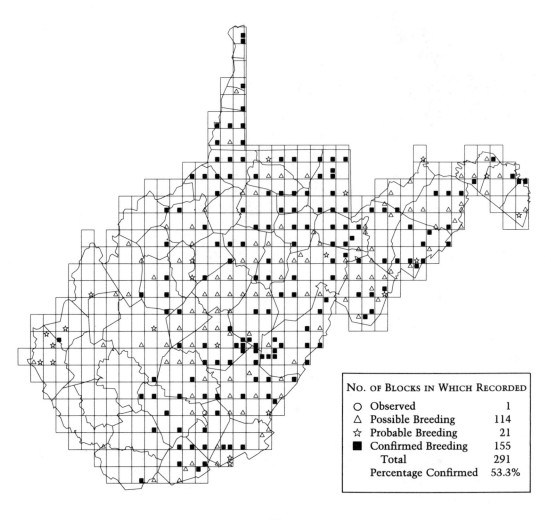

No. of Blocks in Which Recorded	
○ Observed	1
△ Possible Breeding	114
☆ Probable Breeding	21
■ Confirmed Breeding	155
Total	291
Percentage Confirmed	53.3%

Northern Bobwhite *Colinus virginianus*

The whistled *bob white* call of the NORTHERN BOBWHITE was once heard in farm country from southern Ontario and central New England south to Florida and the Gulf Coast. Bobwhite, often called quail, inhabit brushy fields, grasslands, and cultivated land, especially around small farms where diverse crops, hedgerows, and weeds provide good quail habitat. Clean farming methods, conversion from small grains to corn, and loss of farmland have caused a general decline in the Northern Bobwhite population throughout much of its range.

Northern Bobwhite numbers declined through the 1960s and 1970s, and the species was almost eliminated in the northern West Virginia counties by the cold, snowy winters of 1976 through 1978. Although populations are still low in these counties, there are indications that bobwhite are recovering in some places. State BBS routes reported an increase of 73 percent (p <0.01) in bobwhite during the decade of the 1980s.

Hall (1983) placed the largest populations of bobwhite in the Eastern Panhandle, the South Branch of the Potomac River Valley, and the Lower Ohio River and Kanawha River valleys. Atlas workers also found bobwhite in the New River and Meadow River valleys, and along upper Tug Fork. They were scarce from the central and mountain counties north.

No. of Blocks in Which Recorded	
O Observed	3
△ Possible Breeding	79
☆ Probable Breeding	42
■ Confirmed Breeding	34
Total	158
Percentage Confirmed	21.5%

King Rail *Rallus elegans*

The KING RAIL is a very local inhabitant of fresh water swamps in the central and southern United States and on the coastal plain north to New England. West Virginia is on the edge of its range, both in the lower Ohio River Valley and the Eastern Panhandle. Atlas workers reported "confirmed" breeding records for this species in a buttonbush marsh at Green Bottom Wildlife Management Area in Cabell County, and also in marshes near Beech Fork Lake in Wayne County.

Atlasers could have missed this secretive bird in the Eastern Panhandle, where it has nested in the past in Altona Marsh and Albemarle Marsh (Hall 1983). This rail is difficult to locate, and its calls may have been unfamiliar to many West Virginia Atlas workers. King Rail nests are usually well hidden beneath a canopy of cattails or other emergent vegetation on a hummock or in a clump of plants.

The few remaining marshes suitable for breeding of this species need to be protected.

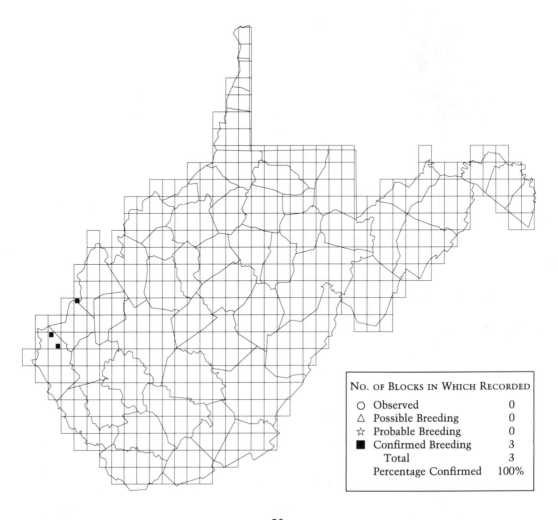

NO. OF BLOCKS IN WHICH RECORDED	
○ Observed	0
△ Possible Breeding	0
☆ Probable Breeding	0
■ Confirmed Breeding	3
Total	3
Percentage Confirmed	100%

Virginia Rail *Rallus limicola*

An elusive bird of cattail and sedge swamps, the VIRGINIA RAIL reaches the southern periphery of its main breeding range in West Virginia. It is more common in central Ohio, northwestern Pennsylvania, and Maryland north to southern Canada.

Atlas records from Tomlinson Run State Park, the North Branch of the Potomac River, Canaan Valley, Green Bottom Wildlife Management Area, and a swamp near Princeton show that this species might be expected in any part of the state where suitable habitat exists. Even small cattail marshes at the inlets of man-made lakes and ponds may harbor a pair.

The Virginia Rail builds its nest in drier areas of the marsh than does the Sora, with which it may share its habitat. The nest is well concealed in marsh vegetation and is difficult to find. The Virginia Rail may be underrepresented on the Atlas map as a result of the bird's secretive habits and some Atlas volunteers' lack of experience with rails.

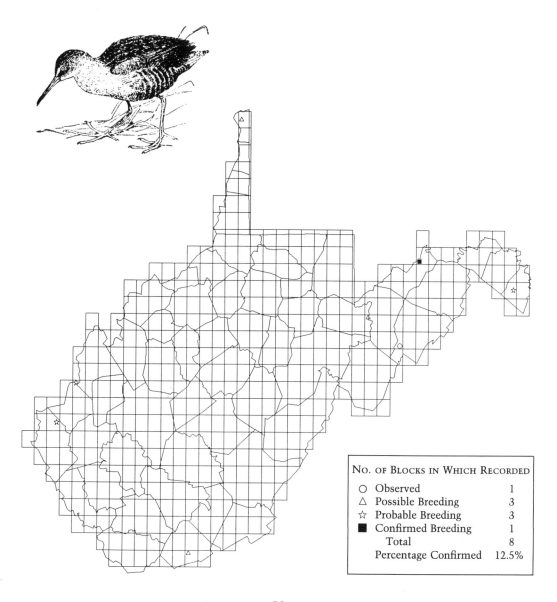

No. of Blocks in Which Recorded	
○ Observed	1
△ Possible Breeding	3
☆ Probable Breeding	3
■ Confirmed Breeding	1
Total	8
Percentage Confirmed	12.5%

Sora *Porzana carolina*

The Sora's loud descending whinny notes and its plaintive *kerwee* whistle are often the only evidence of the bird's presence in its marsh habitat. The nest, made of dead cattail leaves and grasses, is well concealed by overarching vegetation. This and its retiring habits make observation difficult.

West Virginia is on the southern margin of the Sora's eastern range, which extends north into northern Ontario and central Quebec. The Atlas project found it breeding only in the Eastern Panhandle, where both "confirmed" records are from Mineral County. Atlas volunteers found a nest with eggs near Cresaptown, Maryland, and a fledgling from a small marshy pond near New Creek. Soras were also present at Altona Marsh in Jefferson County during the Atlas safe dates.

Migrating Soras were found in swamps in the lower Ohio Valley during the Atlas project, but none remained in that location during the safe dates. As is the case with other rails, the Sora is elusive and difficult to locate. The future of this species in the state depends on preservation and protection of its wetland habitat.

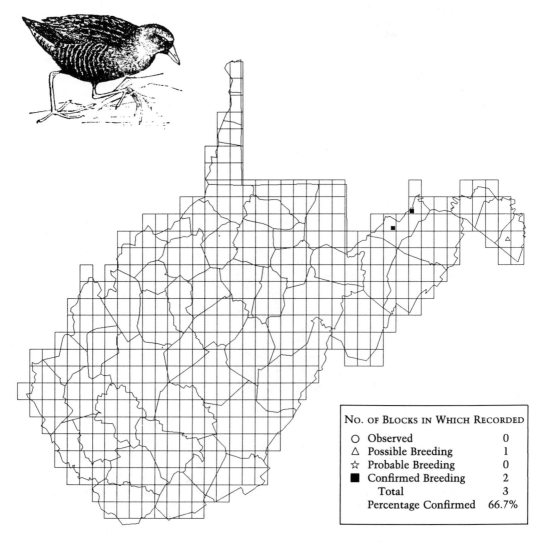

No. of Blocks in Which Recorded	
○ Observed	0
△ Possible Breeding	1
☆ Probable Breeding	0
■ Confirmed Breeding	2
Total	3
Percentage Confirmed	66.7%

Common Moorhen *Gallinula chloropus*

West Virginia is well within the eastern range of the COMMON MOORHEN, which includes all of the eastern United States and southern Ontario and Quebec. Nevertheless, no "confirmed" records of nesting were reported in the state during the Atlas period.

Common Moorhen nests are placed in dense emergent vegetation and may be hard to find, but adults can often be observed swimming in nearby open water. It is doubtful that observant volunteers could have missed this species if it had been present in the breeding season. One would expect it to occur in marshes in Jefferson County, where it has nested in the past (Hall 1983).

Moorhens were seen during the Atlas period at McClintic Wildlife Management Area in Mason County as late as May 15 in 1984 (Igou 1984). A pair, which may have been early fall migrants, was found at Green Bottom Wildlife Management Area in August 1986.

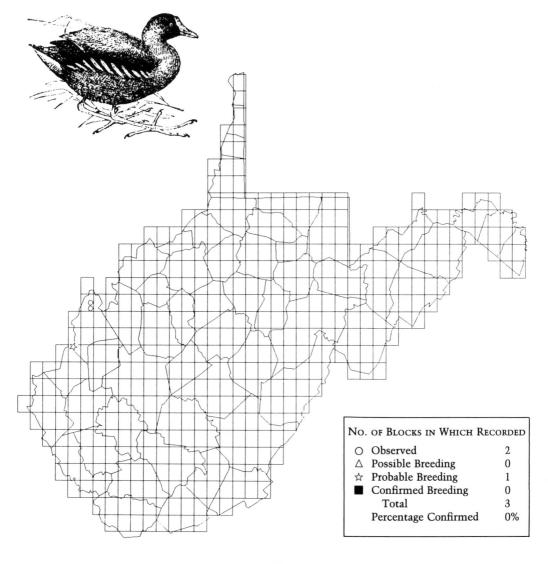

NO. OF BLOCKS IN WHICH RECORDED	
○ Observed	2
△ Possible Breeding	0
☆ Probable Breeding	1
■ Confirmed Breeding	0
Total	3
Percentage Confirmed	0%

Killdeer *Charadrius vociferus*

The KILLDEER is a fairly common bird of open habitats. It has increased in numbers in the central and eastern United States despite a brief population decline after the cold winters of the late 1970s (Robbins, Bystrak, and Geissler 1986). Killdeer increased in West Virginia an average of 6.6 percent per year ($p < 0.01$) between the years 1966 and 1987, according to U.S. Fish and Wildlife BBS data.

This familiar, conspicuous bird often nests near industrial and suburban development, in or near gravel parking lots, sparsely vegetated fields, airports, interstate highway exchanges, golf courses, athletic fields, gravel roads in open country, and even on gravel-covered flat roofs of buildings. It should occur in any county with much unforested land.

Atlas records for Killdeer are scattered in the more heavily forested central Allegheny Mountains. Atlas workers did not record the species in the well-forested Tyler, Lewis, Clay, Lincoln, and McDowell counties. But the Killdeer no doubt occurs in farming and settled areas in Lewis County and other places. The paucity of records in eastern Kanawha County northeast to Lewis County may be a consequence of hasty coverage or of the nature of the Atlas grid rather than an indication of actual low population density. However, this species is conspicuous and easily "confirmed" by presence of fledged young and distraction displays and, although Kanawha County was not well covered, adequate coverage did take place in the counties to the northeast.

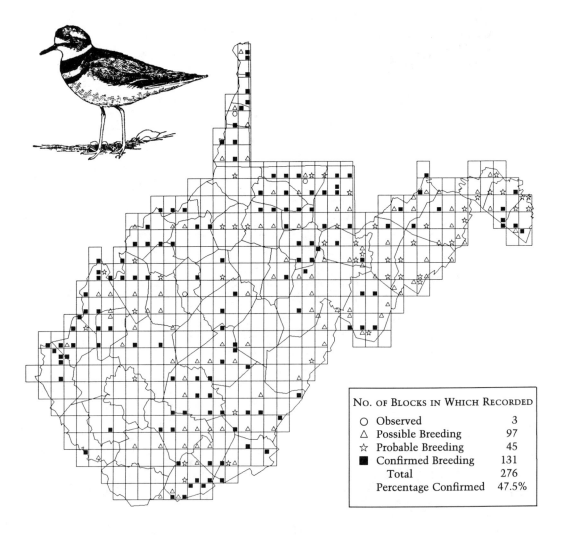

No. of Blocks in Which Recorded	
○ Observed	3
△ Possible Breeding	97
☆ Probable Breeding	45
■ Confirmed Breeding	131
Total	276
Percentage Confirmed	47.5%

Spotted Sandpiper *Actitis macularia*

The SPOTTED SANDPIPER is a conspicuous bird of stream banks in open habitats and farm ponds. The species is fairly common in the northern United States and Canada, from Kentucky and western North Carolina north (AOU 1983). It nests along stream banks and lake shores in open places such as weedy fields or stubble within easy reach of water.

Only three Spotted Sandpiper records were "confirmed." Spotted Sandpipers are late-arriving migrants, and some of the "possible" records may be migrants on their way through the state. The scattered Atlas records reflect a loss of habitat for this species in West Virginia. In recent years, old farmlands have grown up in forest and streams suitable for Spotted Sandpipers have become polluted with acid mine drainage or have been disturbed by recreational use. Fish Creek in Marshall County, for example, where the species bred at one time, has been subjected to severe disturbance from all-terrain vehicles, whose owners drive them in the stream bed and along the banks.

This formerly common species should probably be included on the West Virginia list of vertebrate species of concern. All breeding records should be reported, and streams where the Spotted Sandpiper breeds should be protected from pollution and unnecessary disturbance.

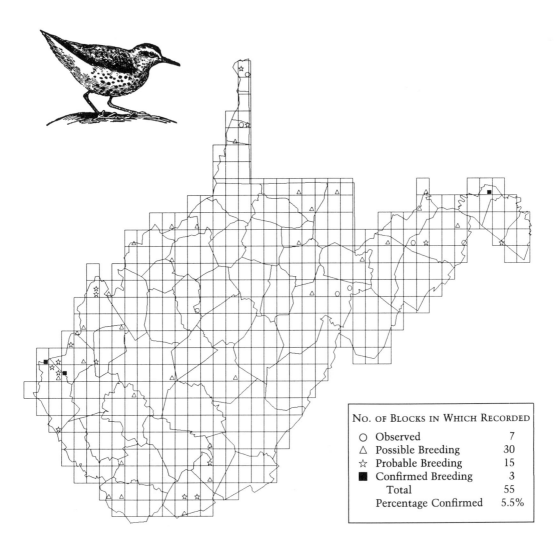

NO. OF BLOCKS IN WHICH RECORDED	
○ Observed	7
△ Possible Breeding	30
☆ Probable Breeding	15
■ Confirmed Breeding	3
Total	55
Percentage Confirmed	5.5%

Upland Sandpiper *Bartramia longicauda*

West Virginia is on the southern margin of the UPLAND SANDPIPER's breeding range, which reaches across the northern United States and southern Canada to central New Jersey and southern Maine (AOU 1983). This species has declined in the East because its grass old-field habitat is disappearing in many places (Tate 1986). Early and frequent cropping of hay may also be a factor in this sandpiper's decline. At the same time, however, reclaimed surface mines in eastern Ohio, West Virginia, and Pennsylvania promise to provide excellent habitat for Upland Sandpipers (Fye 1984). The species is included in the *Vertebrate Species of Concern in West Virginia* (W. Va. DNR n.d.),

and it is considered threatened in Pennsylvania (Gill 1985).

Difficult to locate, Upland Sandpiper nests are built on the ground, well concealed in thick grass, which is arched over the top of the nest. Although Upland Sandpipers were found in only one West Virginia Atlas block (in 1985 and again in 1989 at the same place), several "possible" and "confirmed" records were found not far north and east of the state by the Pennsylvania Atlas (Brauning 1992). This inspires hope that the bird may continue to breed occasionally on grassy mountain tops and reclaimed surface mines in northern West Virginia, where it has nested in the past.

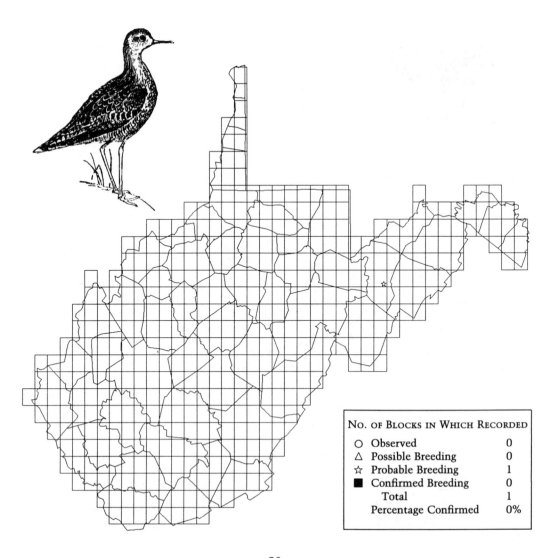

NO. OF BLOCKS IN WHICH RECORDED	
O Observed	0
△ Possible Breeding	0
☆ Probable Breeding	1
■ Confirmed Breeding	0
Total	1
Percentage Confirmed	0%

Common Snipe *Gallinago gallinago*

Breeding range for the COMMON SNIPE extends far into northern Canada and across the northern states. A rare breeder in northeastern Ohio and northern Pennsylvania, it reaches its most southern breeding station in West Virginia's Canaan Valley in Tucker County. A newly fledged snipe chick was found there in 1971 (Hall 1983). Atlas workers found snipe in Canaan Valley in spirea-alder wetlands along the Blackwater River, in Altona Marsh in Jefferson County, and at Mount Storm Lake in Grant County.

Common Snipe are usually found by the sound of their *winnowing* courtship flight. Their nests are very well concealed in vegetation on a clump of raised ground or a partially sunken log in the swamp (Sutton 1923).

Atlas safe dates for Snipe, June 1 through July 10, were set late in the season to avoid confusion with migrants, but breeding birds may have been missed as a result. Our native birds begin their aerial courtship in March, continue through April and May, but display only infrequently in June (Michael 1990).

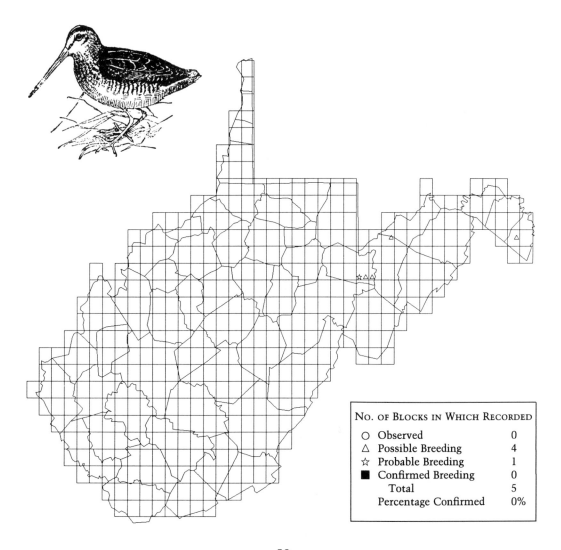

NO. OF BLOCKS IN WHICH RECORDED	
O Observed	0
△ Possible Breeding	4
☆ Probable Breeding	1
■ Confirmed Breeding	0
Total	5
Percentage Confirmed	0%

American Woodcock *Scolopax minor*

The whistled notes and chirps of the AMERI-CAN WOODCOCK's sky dance draw many bird watchers to damp brushy fields and swamps in the late winter and early spring. By early February, many males have returned to southern West Virginia, and nests have been reported as early as late February (Kletzly 1976). The primary breeding range of the woodcock extends across the northeastern United States and southern Canada, south in the Appalachian Mountains through West Virginia. Breeding may occur anywhere east of the Mississippi, from Texas and Florida north.

Woodcock usually nest near the edge of a clearing in alder thickets, brushy fields, young stands of second-growth forest, or along fence rows. The nest is a depression in the ground lined with dead leaves and sur-rounded by a few twigs or grass stems. It is usually placed on a slight elevation near the trunk of a tree, shrub, or clump of vegetation (Kletzly 1976).

West Virginia Atlas volunteers found Woodcock in the Allegheny Mountains Region, the lower Ohio River Valley, and the Northern Panhandle. Most "confirmed" records were of fledged chicks, as the nests are very hard to find without the help of a well-trained dog. The Atlas safe dates, April 10 through September 20, may have caused some volunteers to ignore early spring courtship. Coupled with the presumption that few workers were in the field in the late winter and early spring, the safe dates may have caused Woodcock breeding to be missed in some areas. This is especially true for the central and southern Western Hills Region, where much of the Atlas fieldwork was done by block busting, an activity that usually was scheduled for early summer.

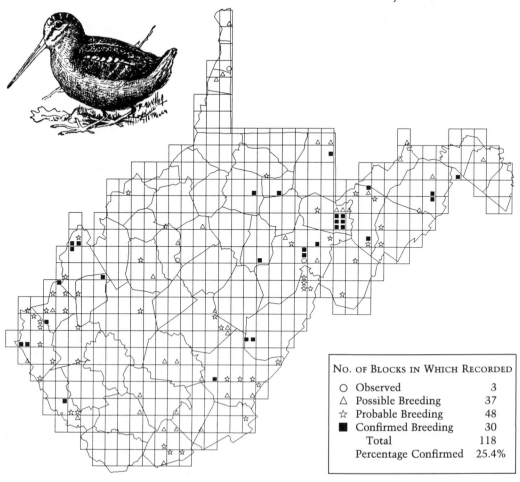

NO. OF BLOCKS IN WHICH RECORDED	
O Observed	3
△ Possible Breeding	37
☆ Probable Breeding	48
■ Confirmed Breeding	30
Total	118
Percentage Confirmed	25.4%

Rock Dove *Columba livia*

The ROCK DOVE, more often called the pigeon, is found throughout the United States and southern Canada. Although it is usually found nesting on city buildings and highway bridges, the species is equally at home nesting in barns, outbuildings, and abandoned houses in rural areas. Less commonly, Rock Doves nest on cliffs. Although some bird watchers tend to ignore pigeons, probably very few occurrences were missed by Atlas volunteers. However, some *X* codes may represent birds that flew far from their roosts to feed.

The decline in agriculture and reforestation of old farms has probably caused a decline in pigeon populations in parts of the state. Campaigns to reduce levels of pigeon "infestations" in some urban areas may also have been responsible. Furthermore, modern architecture is less conducive to pigeon nesting than were Victorian style buildings, many of which have been torn down or remodeled in a cleaner design. Nevertheless, BBS data show pigeon population increases in West Virginia in the years 1965 to 1989. As expected, the Atlas map shows the Rock Dove in all populated areas of the state and in valleys where farming is an important activity.

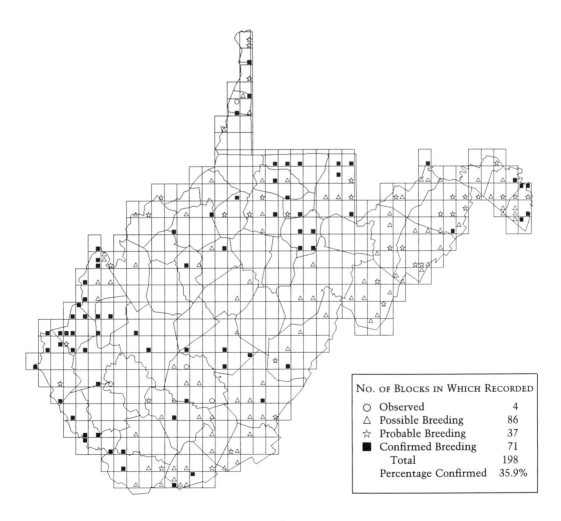

NO. OF BLOCKS IN WHICH RECORDED	
○ Observed	4
△ Possible Breeding	86
☆ Probable Breeding	37
■ Confirmed Breeding	71
Total	198
Percentage Confirmed	35.9%

Mourning Dove *Zenaida macroura*

Found in all 48 contiguous states, the MOURNING DOVE is the most abundant dove in North America. In West Virginia, Mourning Doves may nest as early as February, and they typically raise two to three broods. Thus, this species may be confirmed at any time from late winter to late summer.

The nest, a flimsy-looking platform of sticks, is usually placed on a horizontal branch of a conifer. Mourning Doves favor open habitats with scattered trees. They occupy agricultural areas, suburban yards, parks, orchards, gardens, and small conifer woodlots, and they are often seen along roadsides.

According to Hall (1983), Mourning Doves are scarce in heavily wooded parts of West Virginia. Atlas volunteers confirmed this, finding the species everywhere except in heavily wooded parts of the Allegheny Mountains and Western Hills. There was a significant increase of 13.2 percent ($p < 0.01$) per year in doves reported on BBS routes in West Virginia between 1966 and 1989. Although forests continue to expand and mature in the state, the BBS results might be explained by increased urbanization along roads, even in rural areas, where ornamental conifer plantings and gardens provide good Mourning Dove habitat. The increased winter feeding of birds may also be a factor in the apparent Mourning Dove population increase.

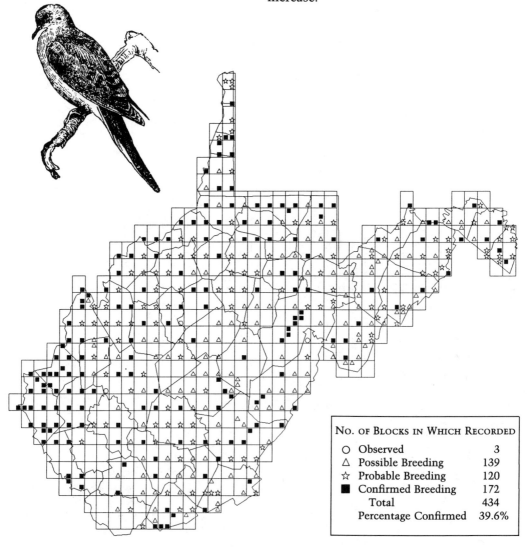

NO. OF BLOCKS IN WHICH RECORDED	
O Observed	3
△ Possible Breeding	139
☆ Probable Breeding	120
■ Confirmed Breeding	172
Total	434
Percentage Confirmed	39.6%

Black-billed Cuckoo *Coccyzus erythropthalmus*

The BLACK-BILLED CUCKOO occurs from east of the Rocky Mountains to the Atlantic Coast and from the Carolinas, northern Alabama, and Tennessee, north to central Canada (AOU 1983). It has a more northern range than the Yellow-billed Cuckoo and is more common than the Yellow-billed Cuckoo at higher elevations. The habitat of the two species is similar except that the Black-billed Cuckoo is found more often in heavier forest and is less common than the Yellow-billed Cuckoo in suburban areas. Cuckoos are secretive, generally silent, and hard to observe. Their nests are usually placed low in a shrub or on a tree branch, well concealed among leaves.

Both species of cuckoos are noted for their role in controlling caterpillars, such as the fall web worm, tent caterpillar, and Gypsy caterpillar. They follow outbreaks of these pests and produce more young in years of heavy infestations. Cuckoos of both species may therefore be much more common in one year than in the next. It is hard to say what impact the sporadic nature of their occurrence might have had on the Atlas effort, but areas that were surveyed in only one year may be less likely to show the presence of cuckoos than those surveyed over several years.

The West Virginia Atlas shows a greater distribution of Black-billed than Yellow-billed Cuckoos in the northern counties and the southwestern portion of the Western Hills Region. The great number of "confirmed" records in the latter area was unexpected, as Kiff et al. (1986) had called the Black-billed Cuckoo a "rare summer resident" in the lower Ohio River Valley.

Although Robbins, Bystrak, and Geissler (1986) reported a significant population increase in the eastern United States, 70 percent ($p < 0.05$) of the BBS routes run in West Virginia between 1966 and 1989 showed a decline in Black-billed Cuckoo populations.

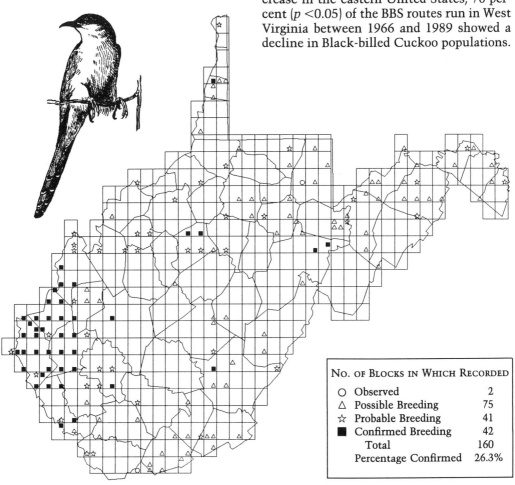

NO. OF BLOCKS IN WHICH RECORDED	
O Observed	2
△ Possible Breeding	75
☆ Probable Breeding	41
■ Confirmed Breeding	42
Total	160
Percentage Confirmed	26.3%

Yellow-billed Cuckoo *Coccyzus americanus*

The YELLOW-BILLED CUCKOO breeds farther south than does the Black-billed Cuckoo, but it has a less extensive northern range, breeding throughout the eastern United States north to southern Ontario and coastal Maine. It is less common in mountainous regions. Much of what has been written here about the behavior of the Black-billed Cuckoo is true also of this species.

As is the case for the Black-billed Cuckoo, the Yellow-billed Cuckoo consumes many hairy caterpillars, web worms, and tent caterpillars. They may be common in areas experiencing heavy infestations of these insects in one year only to become less common when the caterpillars become scarce in succeeding years. Atlas workers who visited blocks for only a year or two may have missed the Yellow-billed Cuckoo as a result of missing a peak year for the caterpillar prey.

As expected, Atlas workers found the Yellow-billed Cuckoo in most well-covered blocks except in the Allegheny Mountains Region. Although Yellow-billed Cuckoos are more likely to occur near human habitation, Atlas workers "confirmed" a lower percentage of Yellow-billed records than of Black-billed. Evidently the Yellow-billed Cuckoo is more widely distributed than the Black-billed Cuckoo, even in forested blocks in the northern Ohio Valley. In 1989, 20 BBS routes reported Black-billed Cuckoos, compared with 24 reporting Yellow-billed Cuckoos.

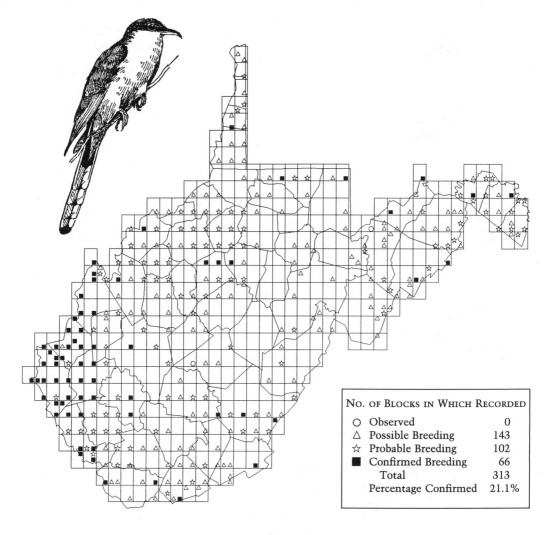

NO. OF BLOCKS IN WHICH RECORDED	
O Observed	0
△ Possible Breeding	143
☆ Probable Breeding	102
■ Confirmed Breeding	66
Total	313
Percentage Confirmed	21.1%

Barn Owl *Tyto alba*

The BARN OWL is a resident of North America north to southern Ontario and southern New England. Prolonged periods of snow cover limit Barn Owl range to the north and in the Appalachian mountains. The National Audubon Society classified the Barn Owl as threatened in 1981 and as a species of special concern in 1986 (Tate and Tate 1982; Tate 1986). It is declining because of habitat loss as agricultural land is converted to forest. Modern farm buildings are tightly constructed and offer few openings for Barn Owl nesting. Collisions with vehicles along interstate highways, which provide excellent hunting habitat on the grassy right-of-ways, may also be an important factor in the decline.

Atlas workers "confirmed" nesting of Barn Owls in the Eastern Panhandle and found scattered records in other parts of the state. Although Atlas records were established in nearby southwestern Pennsylvania (Brauning 1992) and in eastern Ohio (Peterjohn and Rice 1991), the species could not be found in the Northern Panhandle of West Virginia even though it nested there until the late 1970s (Sandercox and Buckelew 1983). As is true of other nocturnal species, the Barn Owl was probably missed by Atlas volunteers in some places. The Atlas project made an effort to overcome this difficulty by advertising for Barn Owl reports in newspapers around the state, which did uncover a few additional records.

The Barn Owl is included on the West Virginia list of vertebrate species of concern (W. Va. DNR n.d.). The West Virginia Nongame Wildlife Program provides Barn Owl nest box plans to people interested in attracting nesting Barn Owls to their property.

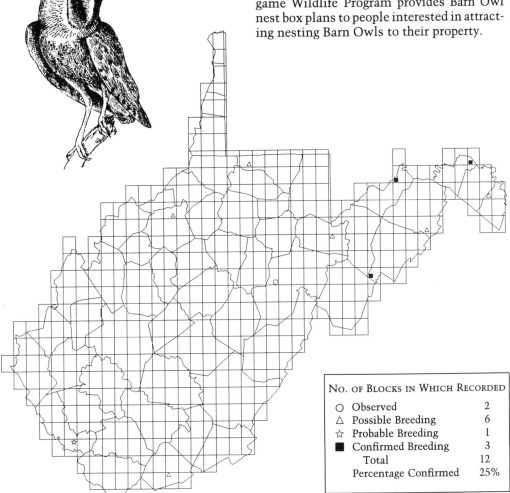

NO. OF BLOCKS IN WHICH RECORDED	
O Observed	2
△ Possible Breeding	6
☆ Probable Breeding	1
■ Confirmed Breeding	3
Total	12
Percentage Confirmed	25%

Eastern Screech-Owl *Otus asio*

The range of the EASTERN SCREECH-OWL in the eastern United States includes all states except those in northern New England. Screech-Owls occur in wooded and suburban areas, where they nest in natural cavities and old Pileated Woodpecker and flicker holes. They also readily use nest boxes.

Although the Eastern Screech-Owl is common in many areas, the National Audubon Society listed it as a species of special concern in 1982 and 1986. Its population is reported to be down in the Appalachian Region (Tate and Tate 1982; Tate 1986), but at least in some parts of West Virginia, it seems to be doing well. For example, this owl is very common in the West Virginia Northern Panhandle. Using taped calls, the Wheeling area Christmas Bird Count reported 154 Screech-Owls in 1987 and 158 in 1988. These were the highest counts in the nation in those years (Monroe 1988; Monroe 1989). Screech-Owl winter territories in the Northern Panhandle are only about one-third of a square mile per pair (Beatty 1988).

The West Virginia and Pennsylvania atlases (Brauning 1992) show a corresponding lack of records in mountainous regions. However, the Eastern Screech-Owl is well distributed in the Ohio and Tug Fork valleys, along the South Branch of the Potomac River and Lost River in the Eastern Panhandle, and in Preston, Marion, and Taylor counties in the north. In parts of the state where volunteers spent considerable time atlasing at night, such as the Northern Panhandle and the Huntington area, the Eastern Screech-Owl was found in almost every priority block. However, since most Atlas survey work was completed during daylight hours, this owl was probably overlooked in some places.

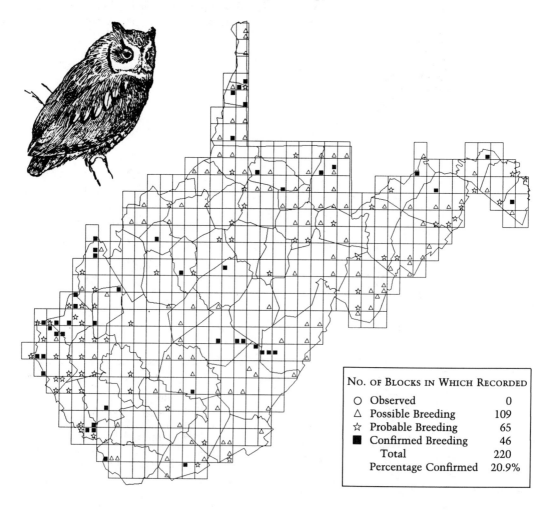

No. of Blocks in Which Recorded	
O Observed	0
△ Possible Breeding	109
☆ Probable Breeding	65
■ Confirmed Breeding	46
Total	220
Percentage Confirmed	20.9%

Great Horned Owl *Bubo virginianus*

The GREAT HORNED OWL is found throughout eastern North America except in the tundra regions of Northern Canada. It probably occurs in every county in West Virginia, but nesting records are few (Hall 1983). No comprehensive population data exist for Great Horned Owls in the state; however, the Great Horned Owl may be increasing because of protective legislation and the public's heightened consciousness of the value of raptors.

This nocturnal raptor prefers extensive woodland. It places its nest on large, abandoned nests of other birds and in natural cavities. Great Horned Owls are more tolerant of fragmented habitat than is the Barred

Owl. These two owls seem to be mutually exclusive, perhaps as a result of the Great Horned Owl's frequent preying on the Barred Owl and the species' differing habitat preferences.

The Great Horned Owl is underrepresented on the Atlas map because most volunteers worked in daylight hours during late spring and summer. Although the species nests in midwinter, calling juveniles are often heard in early summer. Atlas newspaper advertisements asking for Great Horned Owl information helped establish records in some parts of the state, and many records were obtained from people questioned by Atlas field workers.

No. of Blocks in Which Recorded	
○ Observed	4
△ Possible Breeding	61
☆ Probable Breeding	63
■ Confirmed Breeding	23
Total	151
Percentage Confirmed	15.2%

Barred Owl *Strix varia*

The BARRED OWL lives in unbroken tracts of woodland throughout eastern North America. Typically it nests in a tree cavity or in a hollow in the top of a broken trunk, often in dense conifers.

This owl is common in West Virginia's mature deciduous forest, less common in the sparser oak-pine forest of the Ridge and Valley Region, and rare in spruce forest (Hall 1983). The Atlas data show less distribution in the upper Ohio River Valley and central part of the state, probably because of a lack of mature forest. Barred Owls were found most often in the southern Western Hills and Allegheny Mountains regions; along the valleys of the Shenandoah, Potomac, and Lost rivers; and along the South Branch of the Potomac in the Ridge and Valley Region.

Being somewhat less nocturnal than other owls, the Barred Owl can often be heard calling during the day in cloudy weather, and it is therefore easier to locate than the Great Horned Owl. Barred Owls also often call on summer evenings, especially when young owls are present. This owl probably occurs in every county of West Virginia. It is underrepresented on the Atlas map.

NO. OF BLOCKS IN WHICH RECORDED	
O Observed	3
△ Possible Breeding	107
☆ Probable Breeding	100
■ Confirmed Breeding	39
Total	249
Percentage Confirmed	15.7%

Long-eared Owl *Asio otus*

Uncommon in West Virginia, the LONG-EARED OWL reaches the southern limit of its eastern range in the state and in neighboring Virginia. Hall (1983) lists only three breeding records for the species, from Ohio, Tucker, and Mason counties.

This owl is more nocturnal in its habits than are some other owls, and typically it remains well concealed in dense evergreens during the day. Thus it is easily overlooked by observers, even in places where it is more common.

Our only Atlas record for the Long-eared Owl is a brood with at least two fledged young found in a pine woodlot in Nicholas County. Several members of the Brooks Bird Club visited the site during the 1986 Foray at Glenville and "confirmed" the record.

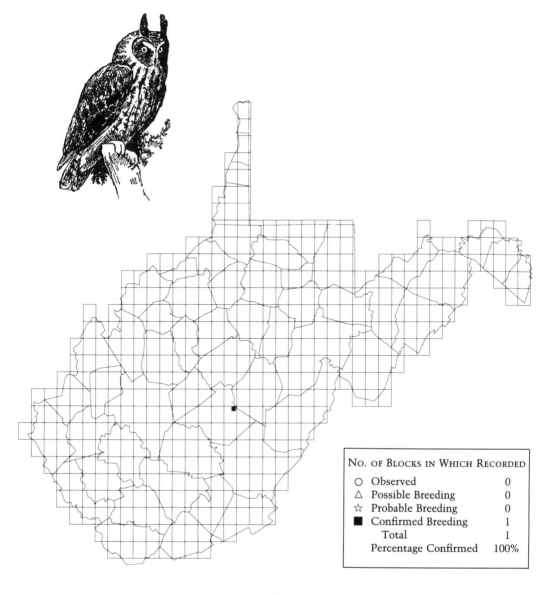

NO. OF BLOCKS IN WHICH RECORDED

○	Observed	0
△	Possible Breeding	0
☆	Probable Breeding	0
■	Confirmed Breeding	1
	Total	1
	Percentage Confirmed	100%

Northern Saw-whet Owl *Aegolius acadicus*

The NORTHERN SAW-WHET OWL is the smallest owl in the eastern United States. It breeds as far south as North Carolina in the spruce belt and mixed spruce-hardwoods. Farther north, the Saw-whet is sometimes associated with deciduous woodlands and edges at lower altitudes (Forbes and Warner 1974). It prefers swampy areas in heavy coniferous forests, where it usually makes its nest in a cavity in a dead tree, although it also will use a nest box. A Saw-whet nest was found in a box erected for northern flying squirrels at Blackbird Knob during the Atlas period.

An exceptional breeding record in the lower Ohio Valley was noted by Edeburn (1968), who collected two juveniles in a small clump of scrub pine in Wayne County. Hall (1983) noted records suggestive of breeding at Morgantown and some late summer reports from Wheeling. Although this species may breed occasionally at low-altitude situations in West Virginia, three out of four Atlas records are from higher altitudes in the spruce belt. The Northern Saw-whet is probably more common in this habitat than Atlas records indicate, but its remote breeding range and strictly nocturnal habits prevent it from being easily observed.

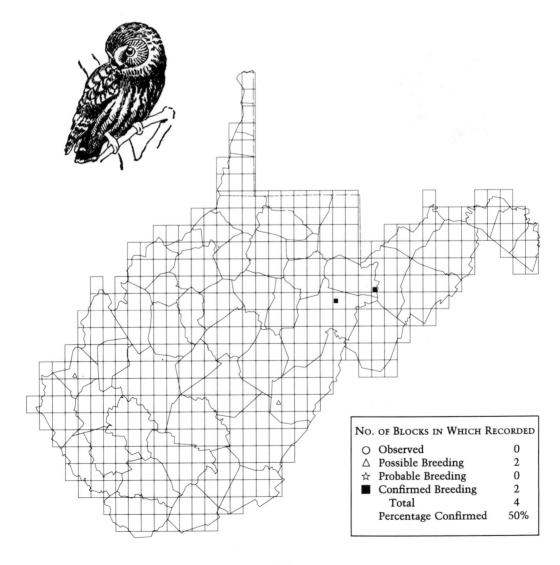

NO. OF BLOCKS IN WHICH RECORDED	
○ Observed	0
△ Possible Breeding	2
☆ Probable Breeding	0
■ Confirmed Breeding	2
Total	4
Percentage Confirmed	50%

Common Nighthawk *Chordeiles minor*

The COMMON NIGHTHAWK breeds from central Canada to Panama. In West Virginia, it nests on flat gravel-covered roofs in towns and cities and on the ground at high elevations in the Allegheny Mountains Region. It may nest on barren ground in large surface mines as well. Atlas volunteers found it in communities throughout the Ohio River Valley; at Morgantown, Charleston, Beckley, Princeton, Elkins; and in towns along the North Branch of the Potomac River as well.

Tate (1986) noted widespread declines in Common Nighthawk populations in the 1970s and 1980s. No such decline has been noted in West Virginia, although there are small cities where no nighthawks were found. During the Atlas period, Brooks Bird Club members searched diligently for night-

hawks, but without success, in Bartow and Durbin in Pocahontas County, Glenville in Gilmer County, Moorefield in Hardy County, and Pineville in Wyoming County. All of these towns have suitable flat-roofed buildings.

The nighthawk is crepuscular, and some Atlas workers trying to cover an area far from home may have failed to cover towns in their block during the early morning and late evening hours. Because one must have access to roofs to find their eggs or young, few nighthawks were "confirmed" during fieldwork.

Because of its history of declining populations, the Common Nighthawk should be watched carefully, and surveys should be organized to monitor populations in West Virginia.

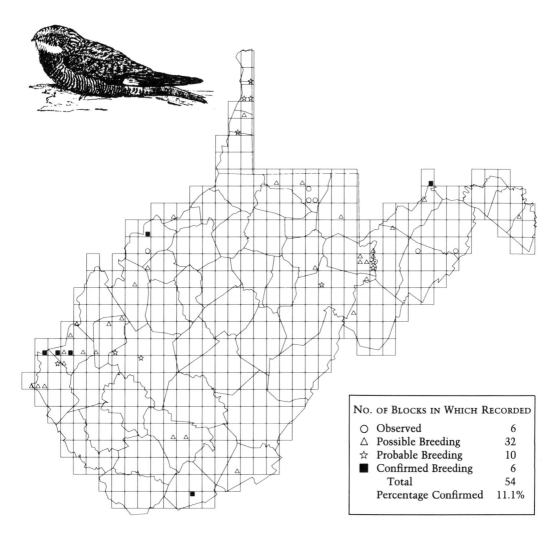

No. of Blocks in Which Recorded	
O Observed	6
△ Possible Breeding	32
☆ Probable Breeding	10
■ Confirmed Breeding	6
Total	54
Percentage Confirmed	11.1%

Chuck-will's-widow *Caprimulgus carolinensis*

The CHUCK-WILL'S-WIDOW is extending its breeding range north into central Pennsylvania (Brauning 1992) and along the coast to Long Island, New York (Meade 1988). In West Virginia, Hall (1983) listed several summer records from central, western, and southern counties and from Berkeley and Hampshire counties in the Eastern Panhandle. There are no nest records for the state, and there were no "confirmed" Atlas records. Of the seven Chuck-will's-widow locations Atlas workers found in West Virginia, all but one (the Morgan County rec-

ord) were in the Western Hills Region.

Chuck-will's-widows are usually found by their distinctive calls. Many observers in West Virginia were unfamiliar with this call; it is therefore probable that some birds went unnoticed or were identified incorrectly as Whip-poor-wills. Two Atlas records of the Chuck-will's-widow were established by sightings of this bird's brilliant golden eye-shine, which shows up in the headlights of an approaching car and is easily distinguished from the Whip-poor-will's red eyeshine.

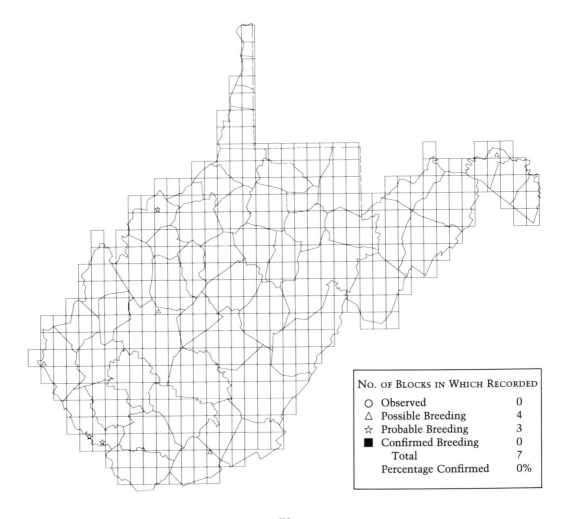

NO. OF BLOCKS IN WHICH RECORDED	
O Observed	0
△ Possible Breeding	4
☆ Probable Breeding	3
■ Confirmed Breeding	0
Total	7
Percentage Confirmed	0%

Whip-poor-will *Caprimulgus vociferus*

In West Virginia, the WHIP-POOR-WILL is rare or absent in many places where it was once common. It is generally agreed that the species has declined throughout the northeastern part of its breeding range (Robbins, Bystrak, and Geissler 1986; Tate 1986), but the reasons for its decline are a mystery. One factor may be a decrease in populations of large moths, which Whip-poor-wills depend on for food. Another may be that Whip-poor-wills are rare in dense forest, and it may be that forests in much of the East have passed through the successional stages that are most hospitable for this species.

In spite of the decline, Whip-poor-wills are still found in every county in West Virginia except Brooke County in the Northern Panhandle and Jefferson County in the Eastern Panhandle. The Atlas project made an intensive effort to find Whip-poor-wills in Brooke County, but even a newspaper campaign failed to reveal any records. The gap in the central Northern Panhandle is real and not attributable to lack of coverage.

Block-busting parties may have missed Whip-poor-wills in blocks where no night coverage was possible, but volunteers frequently were able to ascertain the presence of Whip-poor-wills by questioning local residents. Courtship is rarely observed in this species. There is no true nest; the eggs are placed on dead leaves on well-drained ground.

Only three "confirmed" records of breeding were reported during the Atlas study, and of these, only one was of a nest and eggs. One of the remaining two records was "confirmed" by the presence of fledged young and the other by a distraction display.

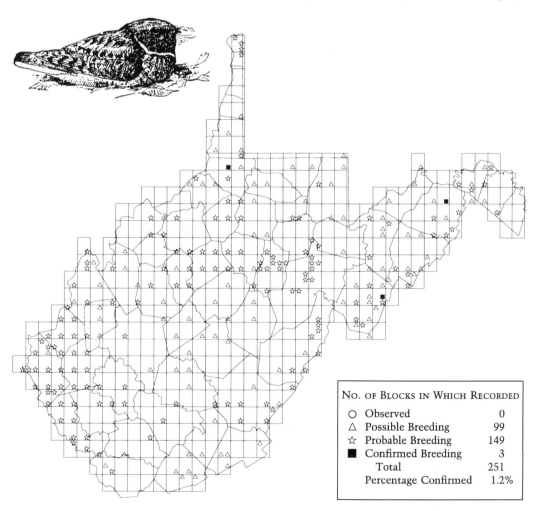

NO. OF BLOCKS IN WHICH RECORDED	
O Observed	0
△ Possible Breeding	99
☆ Probable Breeding	149
■ Confirmed Breeding	3
Total	251
Percentage Confirmed	1.2%

Chimney Swift *Chaetura pelagica*

The CHIMNEY SWIFT breeds from Florida north to central Ontario and Quebec (AOU 1983). It is associated with towns and cities in every West Virginia county. In remote areas, this bird may still nest in hollow trees as it did before the arrival of Europeans, but there are very few records of such nest sites in West Virginia (Hall 1983). Most records are from settled areas, where it can find chimneys and other structures suitable for nest placement. Since nests can rarely be observed directly, most Atlas records were established by observing courtship behaviors, listening for the sound of young birds in chimneys, and other less direct means.

Chimney Swifts were not found in some priority blocks in the more heavily forested Allegheny Mountains and Western Hills regions. Since swifts may feed miles from their nests, it is unclear whether records from remote areas represent birds that breed in natural sites, such as hollow trees, or in the chimneys of distant human settlements.

No. of Blocks in Which Recorded	
O Observed	25
△ Possible Breeding	210
☆ Probable Breeding	66
■ Confirmed Breeding	92
Total	393
Percentage Confirmed	23.4%

Ruby-throated Hummingbird *Archilochus colubris*

The RUBY-THROATED HUMMINGBIRD is the only hummingbird that breeds in the eastern two-thirds of North America, where its range extends north into southern Canada. Populations have been stable throughout its range for the past 25 years (Robbins, Bystrak, and Geissler 1986), but the BBS data show an annual increase of 9.7 percent ($p < 0.01$) in West Virginia during the 1980s. The increase might be due to the large and growing number of people who feed hummingbirds.

The Ruby-throated Hummingbird is common in suburbs and wooded areas containing numerous open patches and edges where flowers are abundant. It is less common in densely wooded forest. In West Virginia, the hummingbird is not common in the Allegheny Mountains Region and is absent from pure spruce forest (Hall 1983). Hummingbirds place their small, lichen-covered nest on a downward slanting branch, 2 to 15 meters from the ground. Viewed from below, the nest looks like a mossy knot on the limb (Harrison 1975).

The Atlas map shows Ruby-throated Hummingbirds in every county and in most priority blocks except in the Allegheny Mountains and in heavily forested areas. Its small nest and rapid flight made nest finding difficult; many "confirmed" breeding records were of fledged young.

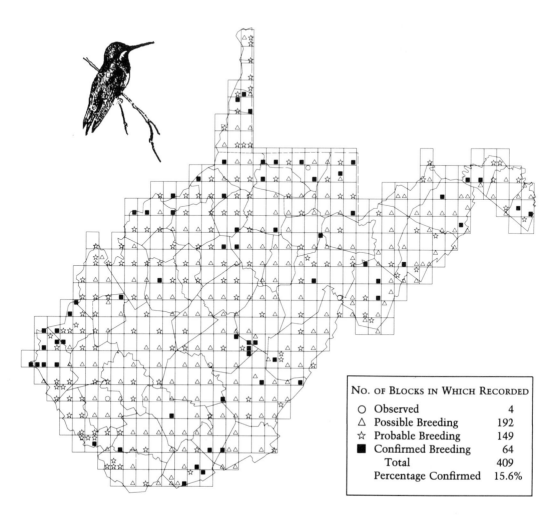

NO. OF BLOCKS IN WHICH RECORDED	
O Observed	4
△ Possible Breeding	192
☆ Probable Breeding	149
■ Confirmed Breeding	64
Total	409
Percentage Confirmed	15.6%

Belted Kingfisher *Ceryle alcyon*

BELTED KINGFISHERS breed from the Gulf Coast north to central Canada. They may range widely along streams and lake shores, but the number of breeding pairs is limited by availability of suitable nesting sites. They usually nest in excavated tunnels in steep stream banks, but they may occasionally use road banks or even tree cavities when stream banks are not available.

Hall (1983) states that many streams that formerly hosted kingfishers in West Virginia are now too polluted to maintain the aquatic life that provides food for this species. Robbins, Bystrak, and Geissler (1986) reported no significant long-term trend in Belted Kingfisher populations, but they noted that the cold winter of 1976–77 caused a temporary decline in its numbers. Atlas volunteers found it to be distributed less extensively in the Allegheny Mountains Region but present in all counties.

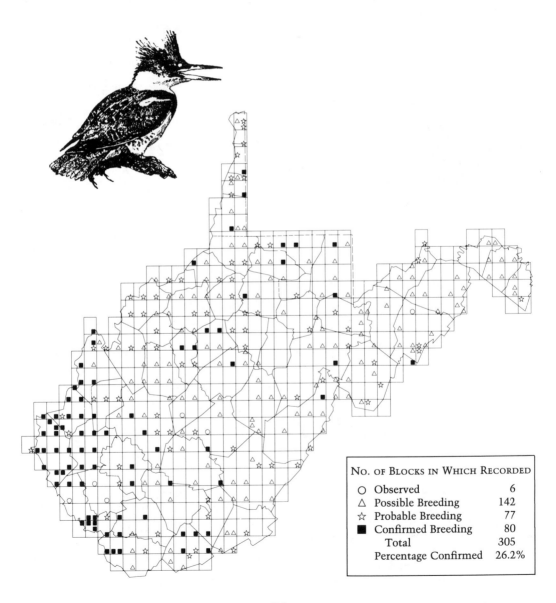

NO. OF BLOCKS IN WHICH RECORDED	
O Observed	6
△ Possible Breeding	142
☆ Probable Breeding	77
■ Confirmed Breeding	80
Total	305
Percentage Confirmed	26.2%

Red-headed Woodpecker *Melanerpes erythrocephalus*

RED-HEADED WOODPECKERS are found locally throughout most of the eastern United States, except for northeastern New England.

This woodpecker subsists almost entirely on mast in the winter, and its summer diet consists of flying insects, grasshoppers, and beetles, supplemented by mast and fruit. Its population decline earlier in this century is attributed to the invasion of the European Starling, which expropriates the Red-head's nest holes. Other possible factors include the chestnut blight that killed off an important source of mast, loss of orchards, and farm abandonment. Collisions with automobiles may take an additional toll. In spite of these problems, Red-headed Woodpecker populations seem to be recovering somewhat in the Northeast in recent years (Robbins, Bystrak, and Geissler 1986).

In West Virginia the species' favored habitat, open oak groves with little understory cover, is scarce (Hall 1983). Atlas volunteers found Red-headed Woodpeckers in the Ohio River Valley and at low altitudes in the Allegheny Mountains Region. Its noisy call note and conspicuous bright, tricolored plumage made it easy to find when present. The Atlas map is probably an accurate representation of Red-headed Woodpecker distribution in the state.

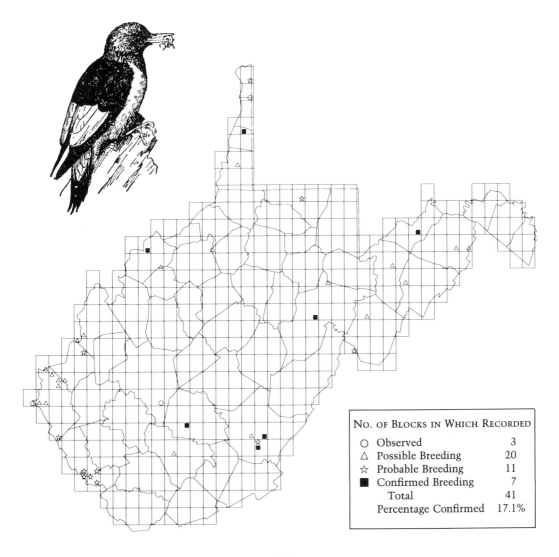

No. of Blocks in Which Recorded	
○ Observed	3
△ Possible Breeding	20
☆ Probable Breeding	11
■ Confirmed Breeding	7
Total	41
Percentage Confirmed	17.1%

Red-bellied Woodpecker *Melanerpes carolinus*

The RED-BELLIED WOODPECKER is a southern species whose breeding range has expanded northward into New York and southern New England in the past century. In West Virginia, it is a fairly common permanent resident (Hall 1983). As the Atlas map shows, it is scarce in the Allegheny Mountains Region except in major river valleys, but it is well distributed elsewhere.

Almost any kind of tree will serve as a Red-bellied Woodpecker nest site, and it occasionally uses an old cavity of a Hairy Woodpecker or reuses one of its own. The European Starling often usurps Red-bellied nest holes, and it has a negative impact on the woodpecker's breeding success. But the Red-bellied Woodpecker is at home in dense forest where starlings are uncommon. This habitat is increasing in West Virginia, and this woodpecker is increasing in numbers. Breeding Bird Survey data show a 9.7 percent ($p < 0.01$) annual median increase in Red-bellied Woodpeckers reported on routes in West Virginia for the years 1980 through 1989.

This colorful woodpecker is well known to bird watchers in the state, and it is a common visitor to feeders. It did not present any difficulty to Atlas volunteers in the field.

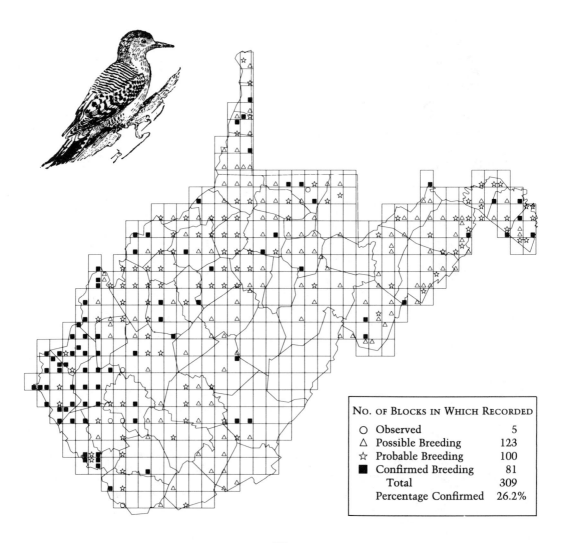

No. of Blocks in Which Recorded	
○ Observed	5
△ Possible Breeding	123
☆ Probable Breeding	100
■ Confirmed Breeding	81
Total	309
Percentage Confirmed	26.2%

Yellow-bellied Sapsucker *Sphyrapicus varius*

The YELLOW-BELLIED SAPSUCKER approaches the southern limits of its breeding range in the Allegheny Mountains Region. According to Hall (1983), this species once nested in fair numbers in the Allegheny Mountains Region with the greatest numbers being in the middle elevations, but it has been decreasing since the 1920s and is now a rare and local summer resident. Hall listed recent nesting records from Randolph and Pocahontas counties. Atlasers found sapsuckers on Burner Mountain and near Buffalo Fork Lake in Pocahontas County, Shaver's Mountain in Randolph County, and Laurel Mountain in Barbour County. The Monongalia County record in a suburban yard in July was probably a nonbreeder or postbreeding individual. Although the Yellow-bellied Sapsucker has spread to lower elevations (some below 150 m) in New York (Levine 1988), there is no reason to believe that this is occurring in West Virginia.

This species nests in remote mountain mixed hardwood-spruce forest in the state. Its noisy behavior around the nest tree and its conspicuous plumage make it easy to locate.

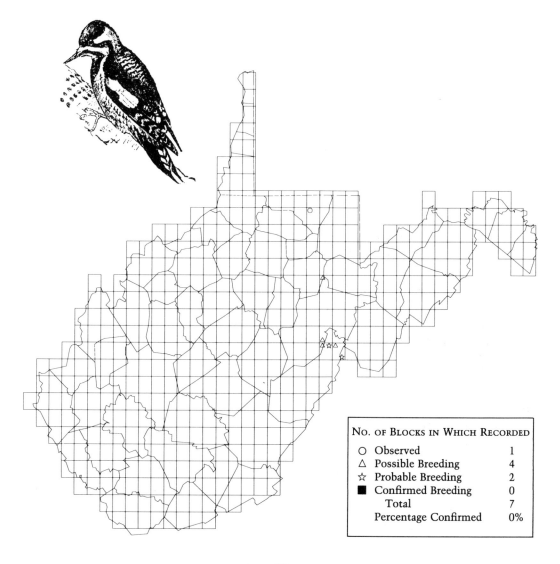

NO. OF BLOCKS IN WHICH RECORDED	
○ Observed	1
△ Possible Breeding	4
☆ Probable Breeding	2
■ Confirmed Breeding	0
Total	7
Percentage Confirmed	0%

Downy Woodpecker *Picoides pubescens*

The Downy, the smallest woodpecker in North America, breeds throughout wooded portions of the eastern United States, achieving its greatest densities clustered in the Appalachians of the Middle Atlantic states (Robbins, Bystrak, and Geissler 1986). Downy Woodpeckers excavate their nest cavities in living or dead trees, often in the underside of a limb. They often re-excavate the same tree year after year.

The Downy Woodpecker inhabits all wooded areas except pure spruce forest and is frequently found in small woodlots or habitats that are not properly considered forest. They are less often found at higher altitudes where the Hairy Woodpecker may be more abundant (Hall 1983).

A frequent visitor to feeders, the Downy Woodpecker is well known to all bird watchers. It is one of the most consistently reported species in the Atlas study. It nests in every county of West Virginia, and it was found in more Atlas blocks than any other woodpecker species.

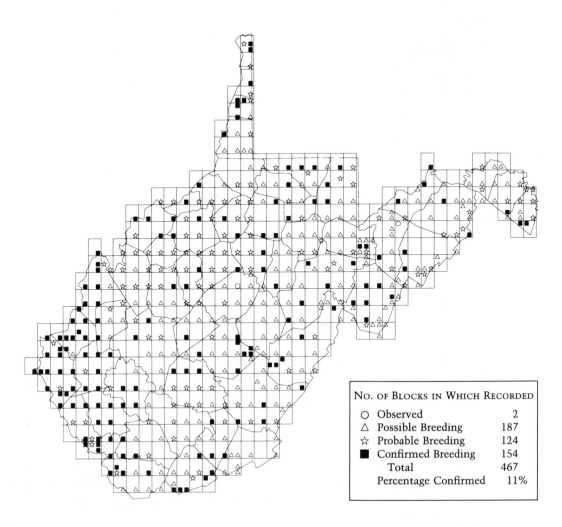

No. of Blocks in Which Recorded	
○ Observed	2
△ Possible Breeding	187
☆ Probable Breeding	124
■ Confirmed Breeding	154
Total	467
Percentage Confirmed	11%

Hairy Woodpecker *Picoides villosus*

The HAIRY WOODPECKER has much the same breeding range as its smaller relative, the Downy Woodpecker. In West Virginia, the Hairy Woodpecker is more common than the Downy Woodpecker at higher elevations and requires more extensive, more mature tracts of forest (Hall 1983).

Atlas workers recorded Hairy Woodpeckers in all counties except Clay. They were found in a larger proportion of priority blocks in the Allegheny Mountains Region and southwestern counties than in other parts of the state. The species may be better distributed in the central part of the state than the Atlas map shows; much of this area was covered by block busting late in the season, when these birds are less conspicuous.

BBS data for the years 1980 through 1989 in West Virginia reveal a significant increase of 3.8 percent ($p < 0.1$) per year in Hairy Woodpecker populations, a reversal of the downward trend seen on BBS routes in earlier decades.

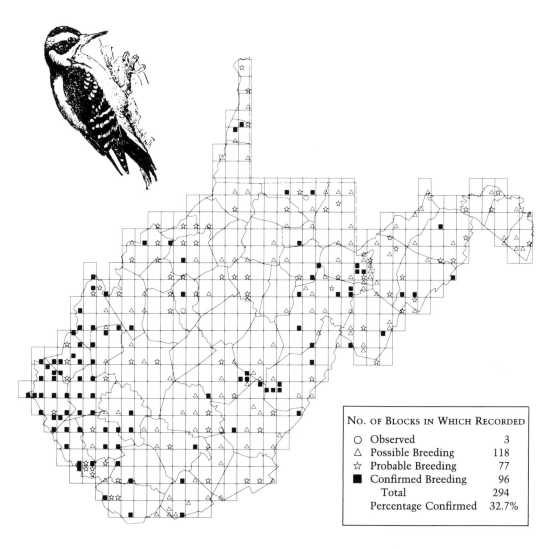

No. of Blocks in Which Recorded	
○ Observed	3
△ Possible Breeding	118
☆ Probable Breeding	77
■ Confirmed Breeding	96
Total	294
Percentage Confirmed	32.7%

Northern Flicker *Colaptes auratus*

The yellow-shafted form of the NORTHERN FLICKER occurs throughout the eastern United States. Its loud *wicka-wicka-wicka* call and persistent drumming is an early sign of spring in West Virginia, where the species is found in every county.

Atlas volunteers failed to note flickers in many blocks in the Western Hills Region away from the Ohio River. The flicker's preference for open habitat makes the bird hard to find in this heavily wooded part of the state. Its inability to compete with the European Starling for nest cavities is considered to be a factor in its general decline in the eastern United States. There was a sig-

nificant decline on West Virginia BBS routes during the years 1966 through 1989 of 4.7 percent per year ($p < 0.01$). The hard winters of 1976 through 1978 also took a toll on this bird.

The Northern Flicker often nests near homes and farms in both suburban and rural areas. The nest hole is often near the top of a high dead stub, and it usually feeds on the ground, often on lawns. Because of these habits, Atlas volunteers had little difficulty finding flicker nests. The rate of "confirmed" records (36%) of breeding for this species was the highest recorded for any of the woodpeckers.

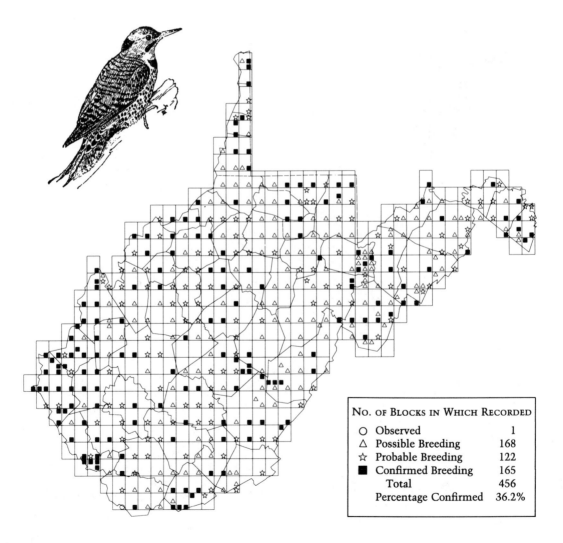

No. of Blocks in Which Recorded	
○ Observed	1
△ Possible Breeding	168
☆ Probable Breeding	122
■ Confirmed Breeding	165
Total	456
Percentage Confirmed	36.2%

Pileated Woodpecker *Dryocopus pileatus*

The PILEATED, North America's largest woodpecker, is a resident from the Gulf Coast and Florida to central Canada. There may be a small southward migratory movement along the Allegheny Front in West Virginia (Hall 1983), but West Virginia Pileated Woodpeckers are thought to be permanent residents.

This woodpecker reaches its greatest distribution in the mixed hardwood forests of the Ohio River Valley, the Western Hills Region, and the Ridge and Valley Region. The Atlas project found it in fewer blocks in the Allegheny Mountains Region. Because of its large size, unmistakable plumage, and characteristic calls, Atlas workers had little trouble finding this conspicuous woodpecker. Its large rectangular feeding holes are diagnostic.

Pileated Woodpeckers are much more common now than in the 1920s and 1930s, when most of the state's forests were in younger second-growth stages (Hall 1983). The species did not occur in the upper Ohio River Valley before 1940 (Buckelew 1976). Breeding Bird Surveys from the 1980s show a high median annual increase in Pileated Woodpecker populations in West Virginia. This trend may be expected to continue as the state's forests mature.

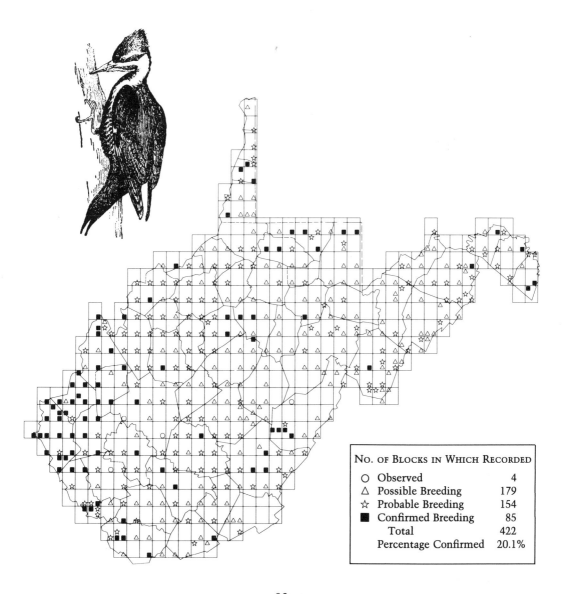

NO. OF BLOCKS IN WHICH RECORDED	
○ Observed	4
△ Possible Breeding	179
☆ Probable Breeding	154
■ Confirmed Breeding	85
Total	422
Percentage Confirmed	20.1%

Olive-sided Flycatcher *Contopus borealis*

The characteristic *pit, per-wheer* or *quick, three beers* call of the OLIVE-SIDED FLY-CATCHER is often heard in openings in northern spruce forest, especially bogs, old beaver ponds, and burned-over slash from lumber operations, where dead standing trees provide singing and feeding perches. This flycatcher has a transcontinental breeding range, which includes much of Canada and the northern United States, with a southern extension along the Appalachians to western North Carolina (AOU 1983). The species is a Neotropical migrant that has declined 5.7 percent ($p < 0.05$) per year on BBS routes in the years 1978 through 1987 (Robbins, Sauer, Greenberg, and Droege 1989). The Olive-sided Flycatcher is listed as a species of special concern in the *Vertebrate Species of Concern in West Virginia* (W. Va. DNR n.d.).

The Olive-sided Flycatcher was once more widely distributed in the mountain counties of West Virginia (Hall 1983). Atlas workers found it only in Randolph and Pocahontas counties, where it could be observed easily during the Atlas period from the boardwalk at Cranberry Glades. The Maryland Atlas project found this flycatcher in the Maryland part of Cranesville Bog, which extends westward into Preston County, West Virginia.

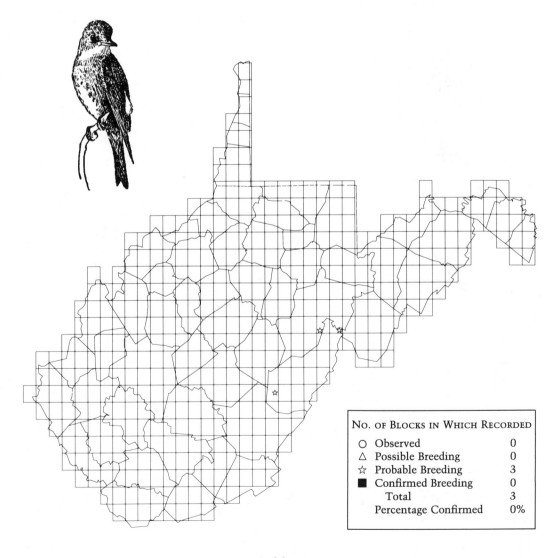

NO. OF BLOCKS IN WHICH RECORDED	
○ Observed	0
△ Possible Breeding	0
☆ Probable Breeding	3
■ Confirmed Breeding	0
Total	3
Percentage Confirmed	0%

Eastern Wood-Pewee *Contopus virens*

The familiar song of the EASTERN WOOD-PEWEE can be heard at any time of the day throughout the summer. Breeding across the eastern United States and southern Canada, this species is strongly associated with oaks and is found in deciduous forest openings, mixed hardwood-coniferous forest, and large shade trees in parks and towns.

The Eastern Wood-Pewee, a Neotropical migrant, has declined 2.1 percent ($p < 0.01$) per year on BBS routes between the years 1966 and 1978 and a lesser amount since then (Robbins, Sauer, Greenberg, and Droege 1989). BBS trends in West Virginia show an average annual decline of 4 percent ($p < 0.01$)

for the years 1966 to 1989. In spite of the noted decline, Eastern Wood-Pewees were most abundant on BBS routes in the southern Piedmont and in Virginia and West Virginia during the years 1966 to 1979 (Robbins, Bystrak, and Geissler 1986).

West Virginia Atlas volunteers found Eastern Wood-Pewees in every county and in most priority blocks except at higher altitudes. Although the species is easy to observe, its lichen-covered nest is well hidden, usually seven or more meters above the ground on a horizontal limb (Harrison 1975). Nest records thus represent a relatively small proportion of Atlas data.

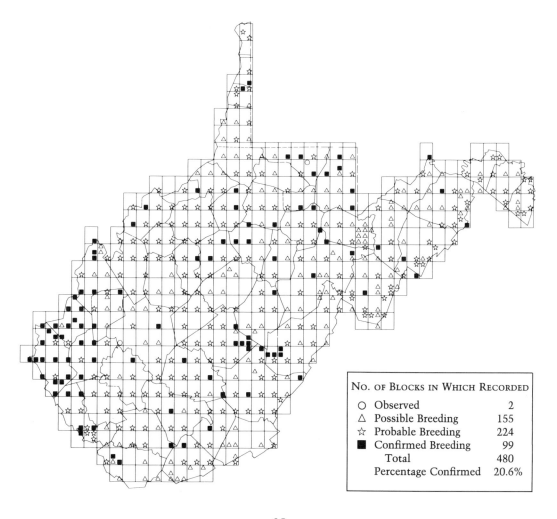

NO. OF BLOCKS IN WHICH RECORDED	
○ Observed	2
△ Possible Breeding	155
☆ Probable Breeding	224
■ Confirmed Breeding	99
Total	480
Percentage Confirmed	20.6%

Yellow-bellied Flycatcher *Empidonax flaviventris*

An inhabitant of boreal forests where the ground is covered with a thick carpet of moss, the YELLOW-BELLIED FLYCATCHER has a main breeding range that is concentrated in Canada. Its range extends south through northern New England and in recent years has been extending sporadically farther south in the mountains to West Virginia and Virginia. The Pennsylvania Atlas found several new breeding locations in that state (Brauning 1992).

A singing male was found in Randolph County, West Virginia, in June 1978, and again at the same place in 1979, 1980, and 1981, but breeding could not be "confirmed" (Hall 1983). Atlas workers discovered the first "confirmed" West Virginia nesting of the Yellow-bellied Flycatcher on June 18, 1988, in the Cranberry Back Country Wilderness of Pocahontas County at an altitude of 1330 meters (Worthington 1989). It was well concealed in a cavity in the root system of an overturned spruce. A "possible" record in Preston County was probably a late migrant, as the Yellow-bellied Flycatcher is the last flycatcher to appear on its breeding range. The Atlas survey may have missed other Yellow-bellied Flycatchers; the species is hard to find in its remote spruce forest habitat, and its calls are easily confused with the calls of the Least Flycatcher and its song with that of the Eastern Wood-Pewee.

Robbins, Bystrak, and Geissler (1986) and Robbins, Sauer, Greenberg, and Droege (1989) reported a significant increase in the species for the northeastern region Breeding Bird Survey. Yellow-bellied Flycatchers increased 14.9 percent ($p < 0.05$) per year between 1966 and 1978 and 3.6 percent since then. This is an exception for the general trend in Neotropical migrants, many of which are decreasing in population.

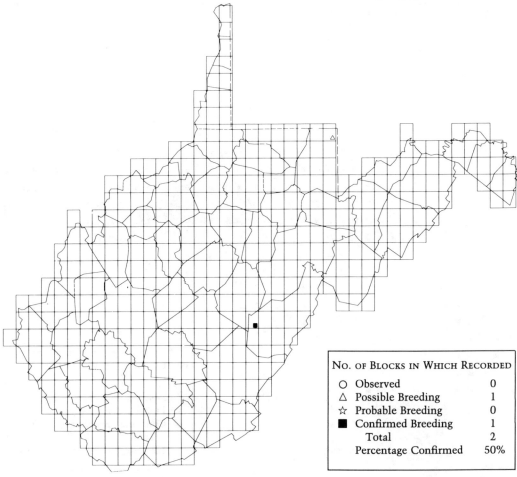

NO. OF BLOCKS IN WHICH RECORDED	
O Observed	0
△ Possible Breeding	1
☆ Probable Breeding	0
■ Confirmed Breeding	1
Total	2
Percentage Confirmed	50%

Acadian Flycatcher *Empidonax virescens*

The ACADIAN FLYCATCHER, one of the confusing *Empidonax* species, is a common woodland bird in West Virginia. Ranging east from the Central Plains to the Atlantic Coast, south to the Gulf Coast, and north to the Great Lakes and southern New England, the Acadian Flycatcher reaches its greatest numbers in Virginia, West Virginia, and the Cumberland Plateau (Robbins, Bystrak, and Geissler 1986). Eastern populations showed slight but significant annual increases of 1.2 percent ($p < 0.05$) from 1966 to 1978 and similar annual declines of 1.4 percent ($p < 0.05$) from 1978 to 1987 (Robbins, Sauer, Greenberg, and Droege 1989).

This flycatcher favors mature mixed deciduous forest that is dissected by small streams and ravines, where it occupies the lower canopy and understory layer. It places its nest near the tip of a lower branch of a large tree, often over water. The nest, with its characteristic long streamers of nest material hanging from the bottom, is relatively easy to spot from below.

Acadian Flycatchers were found in most priority blocks in the Western Hills, and less frequently at higher elevations in the Allegheny Mountains Region and on the dry slopes and oak-pine forests of the Ridge and Valley Region. This is the same distribution described by Hall (1983). Atlas workers found the species in every county of West Virginia.

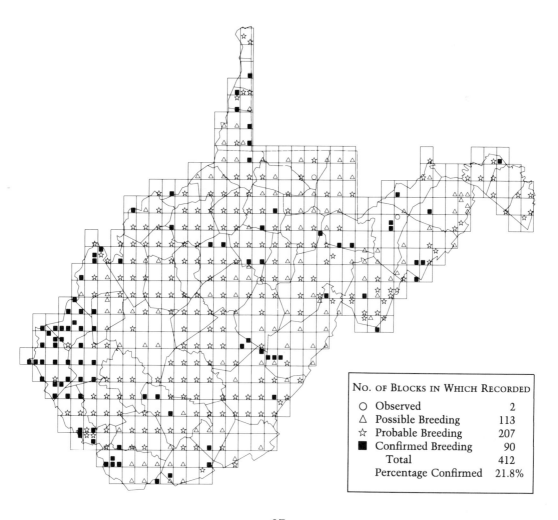

NO. OF BLOCKS IN WHICH RECORDED	
O Observed	2
△ Possible Breeding	113
☆ Probable Breeding	207
■ Confirmed Breeding	90
Total	412
Percentage Confirmed	21.8%

Alder Flycatcher *Empidonax alnorum*

The ALDER FLYCATCHER is a bird of northern alder swamps, reaching south in the Appalachian Mountains to eastern Tennessee (AOU 1983). Hall (1983) called the Alder Flycatcher a fairly common to common summer resident of West Virginia's mountain bogs and areas flooded by beavers. It is more strictly limited to alder-lined marshes than the Willow Flycatcher is to its habitat. The Willow Flycatcher appears to be slowly invading the Alder Flycatcher's range from the west, and it is replacing it in some of its former habitats.

West Virginia Atlas volunteers found the Alder Flycatcher, however, in its mountain habitat in Canaan Valley in Tucker County and in Preston, Randolph, and Pocahontas counties. It was also seen in the southern Allegheny Mountains in Summers and Mer-

cer counties, as well as in Wayne County in the lower Ohio Valley. "Confirmed" records for the species were reported from Jefferson County in the Eastern Panhandle. In Maryland and Pennsylvania, Atlas project workers found the Alder Flycatcher in the mountains that are continuous to the northeast with the Tucker and Preston county populations in West Virginia (Brauning, 1992; Maryland Atlas data).

In the past, the birds now known as the Alder and the Willow flycatchers were considered to be one species, the "Traill's Flycatcher." The two are virtually identical in plumage but differ in nesting habits and song. Most of the Alder Flycatchers in the West Virginia Atlas were identified and differentiated from the Willow by song.

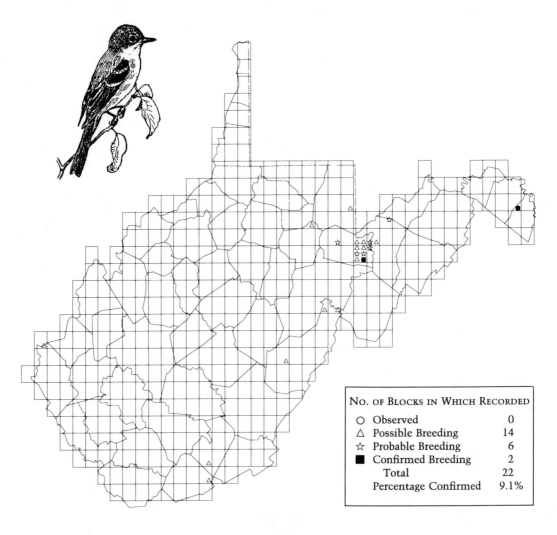

NO. OF BLOCKS IN WHICH RECORDED	
○ Observed	0
△ Possible Breeding	14
☆ Probable Breeding	6
■ Confirmed Breeding	2
Total	22
Percentage Confirmed	9.1%

Willow Flycatcher *Empidonax traillii*

The WILLOW FLYCATCHER is virtually identical in plumage to the Alder Flycatcher, with which it was considered conspecific until 1973 (AOU 1983). The breeding range of the Willow Flycatcher is from southern Canada and central Maine south to southern Ohio and, at higher elevations, through West Virginia, Maryland, and Virginia to Georgia (AOU 1983).

The species nests in willows or alders along stream sides and the edges of marshes and wet meadows, but it is not as strictly limited to this habitat as the Alder Flycatcher is to its alder swamp habitat. Willow Flycatchers are sometimes found in brushy fields and even in suburban sites (Hall 1983). Typically the Willow Flycatcher places its nest one or two meters high in an upright fork or on a horizontal limb of a willow or other shrub with a similar life form. The nest is more compact than the Alder Flycatcher nest, incorporating more cottony plant materials and often, in the rim, feathers.

As in other parts of its range, in West Virginia the Willow Flycatcher sometimes is found close to the Alder Flycatcher, and it may be replacing it in some of its former habitat. The distribution of the Willow Flycatcher as seen on the Atlas map is similar to that described by Hall (1983). It can be found in the Ohio River Valley, valleys of the Allegheny Mountains Region, and locally in the Eastern Panhandle and scattered locations in the Western Hills Region.

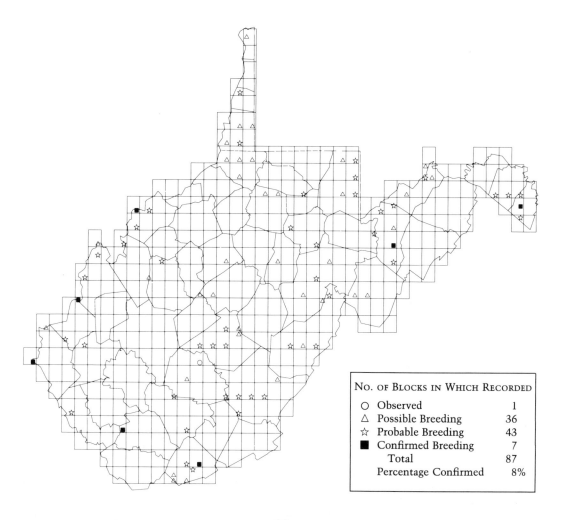

NO. OF BLOCKS IN WHICH RECORDED	
○ Observed	1
△ Possible Breeding	36
☆ Probable Breeding	43
■ Confirmed Breeding	7
Total	87
Percentage Confirmed	8%

Least Flycatcher *Empidonax minimus*

Smallest of the *Empidonax* flycatchers, the LEAST FLYCATCHER is known for its quick, staccato *che-bek* song, which it utters persistently throughout the day during the breeding season. This species avoids the heavy forest and prefers woodland edge, second growth, open woods, orchards, parks, and suburban shade trees. Its breeding range includes southcentral Canada, the northern United States, northern Ohio, and western and northern Pennsylvania, with an extension south in the Appalachian Mountains to Georgia (AOU 1983).

According to Hall (1983), the Least Fly-catcher is found in West Virginia at middle elevations (600 to 900 m) in the Allegheny Mountains Region and western Ridge and Valley Region, and also in the Northern Panhandle. West Virginia populations increased a median 9.5 percent ($p < 0.01$) per year on BBS routes during the 1980s.

Least Flycatchers build a compact nest in the crotch of a limb or on a horizontal limb in almost any kind of tree, typically 3 to 6 meters from the ground. Atlas volunteers found only two nests; most "confirmed" records were of fledged young or adults with food for young.

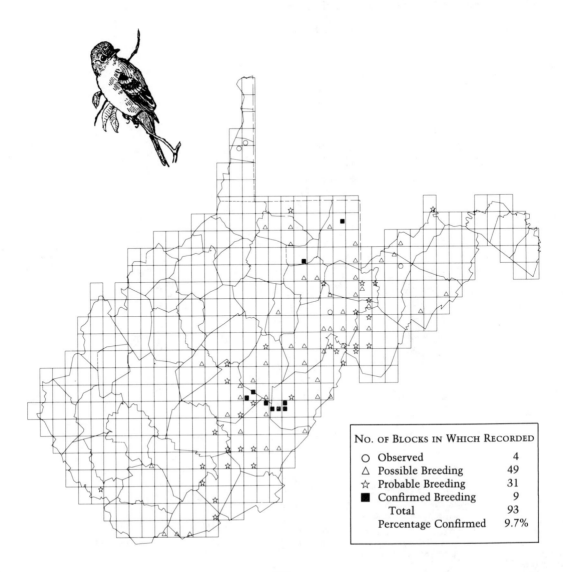

NO. OF BLOCKS IN WHICH RECORDED	
○ Observed	4
△ Possible Breeding	49
☆ Probable Breeding	31
■ Confirmed Breeding	9
Total	93
Percentage Confirmed	9.7%

Eastern Phoebe *Sayornis phoebe*

Now the most widely distributed flycatcher in West Virginia, the EASTERN PHOEBE was probably only locally distributed in pre-settlement times. At that time, it was restricted to open woodlands bordering streams where rocky cliffs with suitable nest sites existed. Although phoebes still use this habitat, they rapidly adapted to the presence of Europeans, using outbuildings, bridges, and other structures as nest sites. The Eastern Phoebe's breeding range extends from central and eastern Canada south to Texas, Louisiana, Alabama, Georgia, and western parts of the Carolinas (AOU 1983).

Populations appeared stable until the severe winter of 1976–77. After that winter, the species began to recover, but the long-term trend is still a significant decline (Robbins, Bystrak, and Geissler 1986). There has been a slight decline on BBS routes in West Virginia in recent years.

The Atlas project "confirmed" phoebes in most priority blocks in every county in West Virginia. The species was absent from some lightly covered blocks in the Allegheny Mountains Region, where it may be somewhat less common. Phoebes are generally absent from pure spruce forest.

The Eastern Phoebe nests were very easy to find because of their location on buildings and bridges. The characteristic moss-covered nest is distinctive, and many phoebes were "confirmed" by their used nests. The Eastern Phoebe is the thirteenth most frequently reported species in the Atlas data.

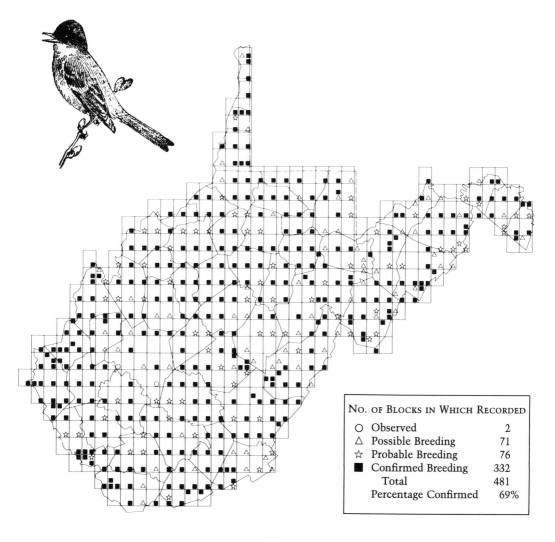

No. of Blocks in Which Recorded	
○ Observed	2
△ Possible Breeding	71
☆ Probable Breeding	76
■ Confirmed Breeding	332
Total	481
Percentage Confirmed	69%

Great Crested Flycatcher *Myiarchus crinitus*

The GREAT CRESTED FLYCATCHER nests from southern Canada south to the Gulf Coast and southern Florida. It prefers relatively mature oak and mixed deciduous forest, but it will inhabit second-growth forest if enough old trees remain for nest cavities. Great Crested Flycatchers prefer natural cavities, even when the cavity is much larger than needed.

Atlas volunteers identified most of these flycatchers by their loud calls. Areas most deficient in Great Crested Flycatcher records (from Kanawha County south and Webster County and western Randolph County north through Harrison County) are also the areas with the lightest Atlas coverage. Block busting accounted for much of the coverage in these counties, and many Great Crested Flycatchers could have been missed because they are quiet and inconspicuous later in the breeding season, when much of the Atlas work was performed.

The Atlas data confirm Hall's (1983) statement that these birds are scarce in northern hardwoods and absent in pure spruce forest.

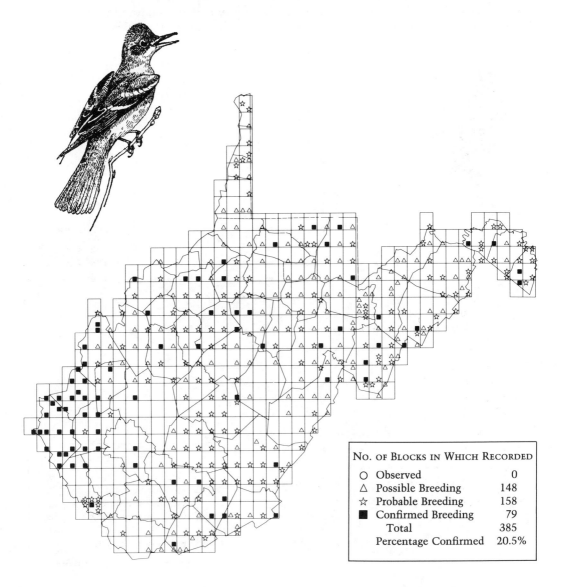

No. of Blocks in Which Recorded	
O Observed	0
△ Possible Breeding	148
☆ Probable Breeding	158
■ Confirmed Breeding	79
Total	385
Percentage Confirmed	20.5%

Eastern Kingbird *Tyrannus tyrannus*

The EASTERN KINGBIRD's vigorous defense of its territory makes it a very conspicuous bird. Eastern Kingbirds occupy open country with tall isolated trees or rows of trees. The species also breeds in open woodlands and orchards and along swamp edges, lake and river shores, and roadsides. It is found during the breeding season throughout central and eastern Canada south to the Gulf Coast and southern Florida.

The large and bulky nest of this species is often conspicuous, typically placed on a horizontal limb far from the trunk and often over water. Sometimes the nest is built in an unusual site, such as on a fence post,

street lamp, telephone pole, or building (Harrison 1975).

According to Hall (1983) the Eastern Kingbird is a summer resident throughout the state, becoming scarce in high mountain valleys, and local and uncommon in much of heavily forested central and southern West Virginia. Atlas workers found the species in every county except Clay. It was recorded in most priority blocks in the Ohio Valley, the northern parts of the Western Hills Region, the Ridge and Valley country of the Eastern Panhandle, and the southeastern counties. Atlas volunteers found it relatively easy to "confirm."

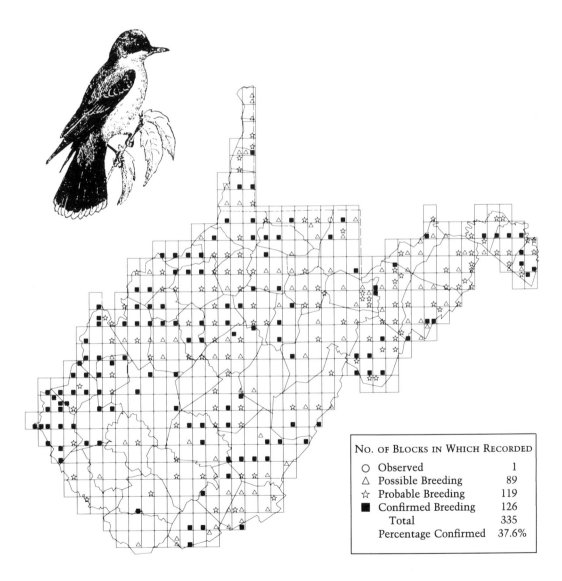

NO. OF BLOCKS IN WHICH RECORDED	
○ Observed	1
△ Possible Breeding	89
☆ Probable Breeding	119
■ Confirmed Breeding	126
Total	335
Percentage Confirmed	37.6%

Horned Lark *Eremophila alpestris*

The aerial courtship song of the HORNED LARK can be heard over its open grassland habitat from the tundra of Canada south to southern Tennessee, northern Alabama, and North Carolina (AOU 1983). Its typical habitat in West Virginia includes short grass pastures, airports, parks, golf courses, athletic fields, old surface mines, and grassy mountaintops (Hall 1983).

Horned Larks place their nests in shallow depressions in the ground, usually next to a tuft of grass or weeds. None of these well-concealed nests was found by West Virginia Atlas workers. All "confirmed" breeding records were either of fledged young or adults with food for young. Most records were established by listening for the Horned Lark's courtship song in suitable habitat.

The Horned Lark declined on West Virginia BBS routes between 1966 and 1989 by a median 4.3 percent ($p <0.1$) per year. Only 3 BBS routes reported Horned Larks during the 1980s, compared with 16 reporting for the years 1966 through 1989. The Atlas map shows Horned Larks in the Northern Panhandle, the lower Ohio River Valley east of Huntington, and scattered places in the Allegheny Mountains and Ridge and Valley regions. A general decline of the species in the Northeast is attributed to abandonment of farms and changes in farming practices.

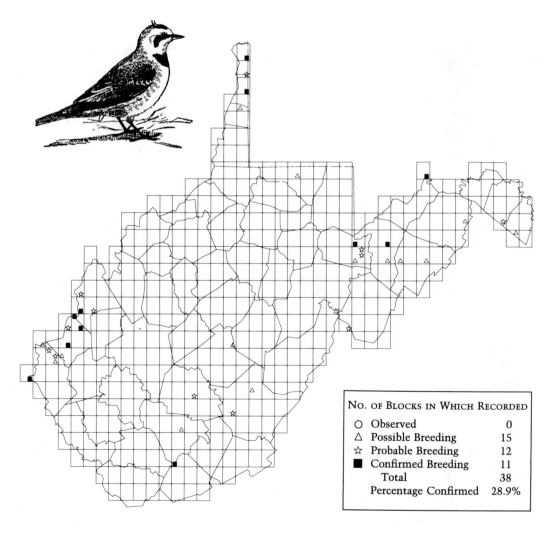

No. of Blocks in Which Recorded	
○ Observed	0
△ Possible Breeding	15
☆ Probable Breeding	12
■ Confirmed Breeding	11
Total	38
Percentage Confirmed	28.9%

Purple Martin *Progne subis*

The PURPLE MARTIN has been closely associated with human settlements for centuries. Populations of Purple Martins were greater than at present in eastern North America until the House Sparrow and European Starling arrived on the scene. Since then, competition for nest boxes from these alien species has caused a large decline in martins. In recent decades, occasional cold, wet spring or summer weather has caused further declines.

Hall (1983) believed that Purple Martins may nest in every county; he noted, however, that they do not occur at high elevations. The West Virginia Atlas project failed to find the species in most of the Allegheny Mountains Region and in several counties of the Western Hills Region, where it was only locally distributed. Atlas workers discovered colonies in six blocks in the three easternmost Eastern Panhandle counties. The bird's close association with human settlement make Purple Martins easy to find. Most breeding records were "confirmed" by observing birds entering and leaving nest boxes.

Survival of this bird is highly dependent on human cooperation because most Purple Martins nest in large nest boxes erected for this purpose. Tate (1986) listed the Purple Martin as a species of special concern and reported decreasing populations in the Appalachian Region. According to Tate, "It is past time to organize a nest box and nest predator management program. This is an easily manipulated species that should not be on this list."

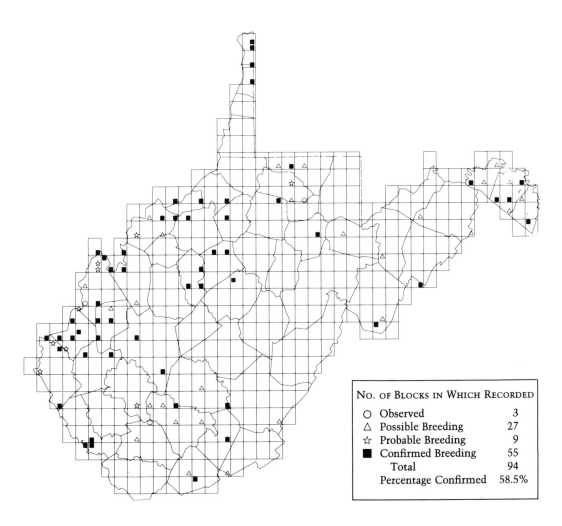

NO. OF BLOCKS IN WHICH RECORDED	
○ Observed	3
△ Possible Breeding	27
☆ Probable Breeding	9
■ Confirmed Breeding	55
Total	94
Percentage Confirmed	58.5%

Tree Swallow *Tachycineta bicolor*

The TREE SWALLOW breeding range extends far north into Canada to northern Ontario, Quebec, and central Labrador. It approaches the southern limit of its principal eastern breeding range in southern West Virginia, western Virginia, and eastern Kentucky (AOU 1983).

Tree Swallows inhabit open situations near water, especially favoring beaver ponds and other flooded places where trees killed by rising water provide plentiful nest cavities. In recent decades, Tree Swallows have expanded their habitat into fields and other open places where nest cavities were not available until the erection of many nest boxes for Eastern Bluebirds. Robbins, By-strak, and Geissler (1986) reported increases of Tree Swallow populations in the eastern region for the years 1965 and 1979. BBS routes in West Virginia have reported an annual median increase of 2.4 percent ($p < 0.01$) for the years 1966 through 1989.

Atlas workers recorded most Tree Swallows in the Allegheny Mountains and Ridge and Valley regions. They also found them in the lower Ohio River Valley, and there were scattered records in the central and northern Western Hills Region. Because of their active, noisy flight over open habitats, Tree Swallows are not easily missed. Nests are often nearby, so a relatively high proportion (46.5%) of Atlas records were "confirmed."

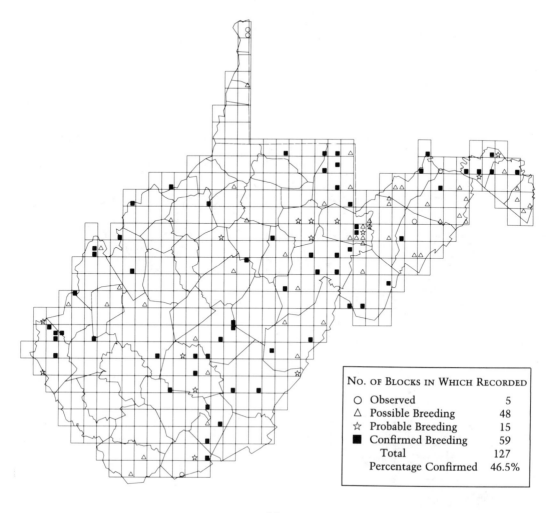

NO. OF BLOCKS IN WHICH RECORDED	
○ Observed	5
△ Possible Breeding	48
☆ Probable Breeding	15
■ Confirmed Breeding	59
Total	127
Percentage Confirmed	46.5%

Northern Rough-winged Swallow
Stelgidopteryx serripennis

The NORTHERN ROUGH-WINGED SWALLOW nests in solitary pairs or small groups in old kingfisher burrows, drainpipes, and other holes under bridges, and in retaining walls, spillways, and abandoned rodent holes. Occasionally the species nests in Bank Swallow colonies.

The Rough-winged Swallow has expanded its breeding range northward over the past several decades into southern Canada, and the BBS data in both eastern and central regions showed significant increases during the period 1966 through 1979 (Robbins, Bystrak, and Geissler 1986). The reason for this species' northward movement is unknown. Unlike some other species, such as the Tufted Titmouse and Red-bellied Wood-

pecker, which have expanded their ranges north, the Northern Rough-winged Swallow is not a permanent resident in its northern range and is not attracted to feeders.

West Virginia Atlas volunteers encountered Northern Rough-winged Swallows in relatively few blocks in the Allegheny Mountains Region and found it most widely distributed in southern portions of the Western Hills Region. This distribution also continues in nearby Ohio and Pennsylvania (Peterjohn and Rice 1991; Brauning 1992).

Some volunteers may have either missed this species or confused it with the less common Bank Swallow. Rough-winged Swallows' dull plumage makes them hard to distinguish on the wing.

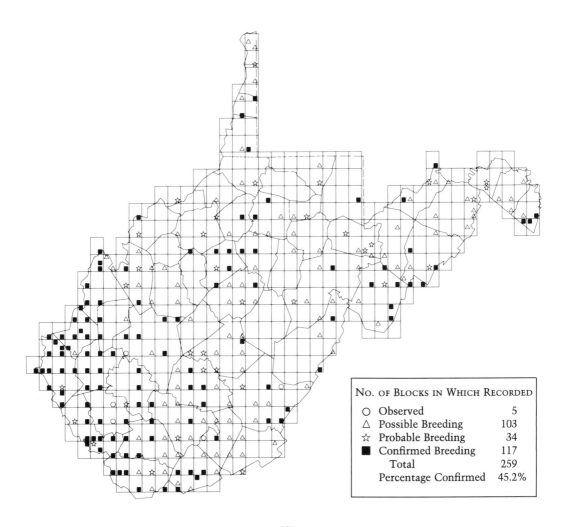

NO. OF BLOCKS IN WHICH RECORDED	
O Observed	5
△ Possible Breeding	103
☆ Probable Breeding	34
■ Confirmed Breeding	117
Total	259
Percentage Confirmed	45.2%

Bank Swallow *Riparia riparia*

The BANK SWALLOW is a colonial nester. It places its nests at the ends of tunnels burrowed near the top of steep erosion banks of streams, gravel or sand pits, and surface-mine high walls. A Holarctic species, the Bank Swallow's breeding range in the East extends from northern Ontario, central Quebec, and southern Labrador south to Arkansas, Tennessee, northern Alabama, central West Virginia, eastern Virginia, and casually south to the Carolinas (AOU 1983).

Bank Swallows are extremely local breeders in West Virginia, and their nesting habitat is transient. Colonies in eroded stream banks usually persist for only a few years. Colonies in gravel and sand quarries and similar habitats are dependent on human activities. Hall (1983) had nesting records for Jefferson, Berkeley, Hampshire, and Mineral counties in the Eastern Panhandle.

Atlas workers found no colonies in these eastern counties but did find colonies in Mason, Jackson, Mingo, Raleigh, Monroe, and Grant counties, where Hall had no breeding records. The transient nature of Bank Swallow colonies is also evident in the significant ($p < 0.01$) decrease of about 2.2 percent per year in this species for the years 1966 through 1987 on the six BBS routes reporting this swallow in the state, and in the 7.1 percent increase per year reported for the 1980s on five routes.

Bank Swallows may forage several kilometers from their colonies and thus may have been recorded as "possible" breeders in blocks in which there are no colonies. The Atlas map shows "possible" records in several blocks near blocks in which breeding was "confirmed."

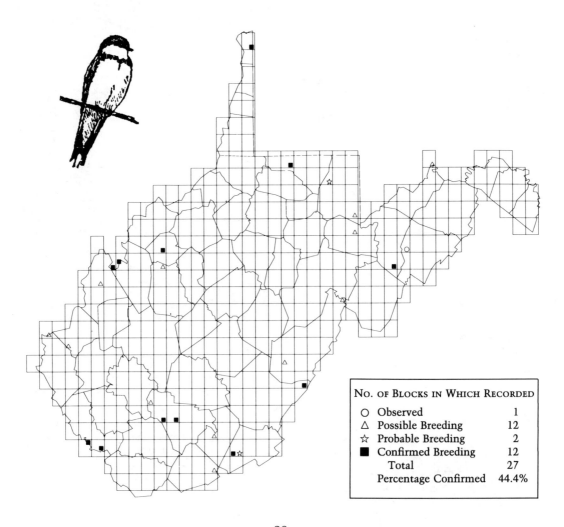

NO. OF BLOCKS IN WHICH RECORDED	
O Observed	1
△ Possible Breeding	12
☆ Probable Breeding	2
■ Confirmed Breeding	12
Total	27
Percentage Confirmed	44.4%

Cliff Swallow *Hirundo pyrrhonota*

CLIFF SWALLOWS build their flask-shaped, mud nests under the eaves of buildings and on bridges, dams, and cliffs in open situations near water. They breed from northern Ontario, southern Quebec, and New Brunswick south to northern parts of the Gulf states, western South Carolina, and southern North Carolina (AOU 1983).

Cliff Swallow populations declined in the first half of the century, perhaps because of competition with House Sparrows and a decline in unpainted barns to which they could attach their nests (Robbins, Bystrak, and Geissler 1986; Gross 1942). West Virginia Cliff Swallows often build their nests just inside the open doors of barns and sheds, commonly attaching their nest to unpainted beams (Samuel 1969; 1971). The population decline has reversed as Cliff Swallows have adapted to building their nests on large dams and highway bridges, and much of the former breeding range is now reoccupied (Robbins, Bystrak, and Geissler 1986).

West Virginia Atlas workers found nesting colonies in the northern Allegheny Mountains Region, the Eastern Panhandle, and in eight counties—Pleasants, Jackson, Putnam, Cabel, Lincoln, Wayne, Mingo, and Mercer—in the Western Hills. Nesting colonies consisted of only a few nests. A colony of 16 on a bridge girder at Beech Fork State Park was typical (Igou 1986). The conspicuous nesting habits of this species allowed Atlas workers to "confirm" breeding in a large number of the records reported.

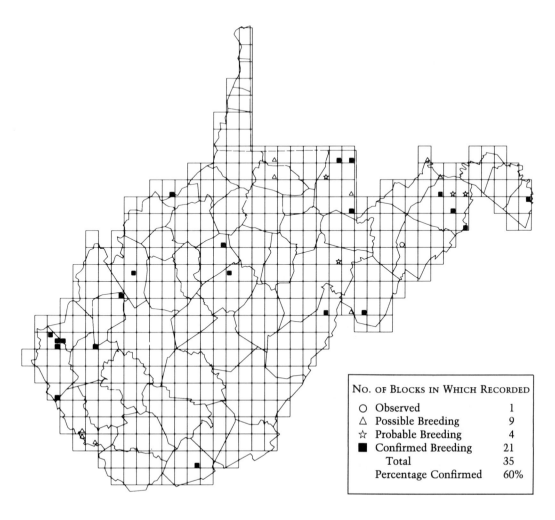

NO. OF BLOCKS IN WHICH RECORDED	
○ Observed	1
△ Possible Breeding	9
☆ Probable Breeding	4
■ Confirmed Breeding	21
Total	35
Percentage Confirmed	60%

Barn Swallow *Hirundo rustica*

The BARN SWALLOW is a familiar bird of open rural areas. Its range extends from northern Ontario, southern Quebec, and Newfoundland, to the Gulf Coast and northern Florida in the south (AOU 1983).

This species nests in barns and outbuildings, under the roofs of porches, under bridges, and in culverts. They still occasionally build their nests inside the entrances of caves, such as the Sinks of Gandy in Pocahontas County, as they must have in preset-tlement times. The nest, made of mud and straw and usually attached to beams inside buildings, is easy to find.

The Barn Swallow is the most common swallow in West Virginia (Hall 1983). Atlas workers found it in every county, but a few blocks in more heavily forested parts of the Western Hills and Allegheny Mountains regions had no Barn Swallows. Most records (73%) were "confirmed."

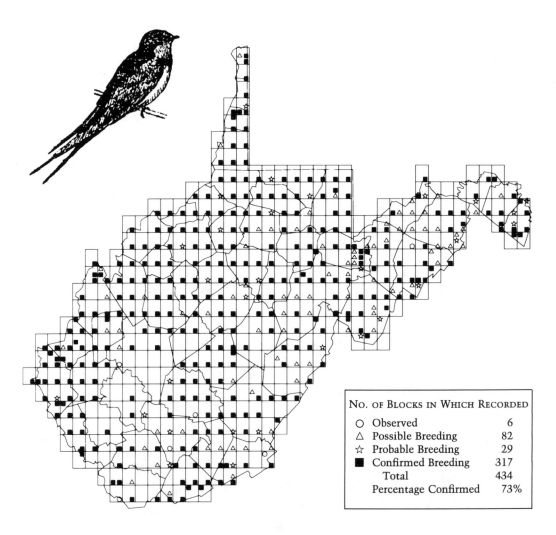

No. of Blocks in Which Recorded	
O Observed	6
△ Possible Breeding	82
☆ Probable Breeding	29
■ Confirmed Breeding	317
Total	434
Percentage Confirmed	73%

Blue Jay *Cyanocitta cristata*

The colorful, noisy BLUE JAY is found throughout the eastern United States and southern Ontario, Quebec, and Newfoundland. Uncommon in West Virginia in the first half of the century, the Blue Jay is now widespread and fairly common to common (Hall 1983). The U.S. Fish and Wildlife Service Breeding Bird Survey in West Virginia has had an average increase of 3.3 percent ($p < 0.05$) per year in Blue Jays counted for the years 1966 to 1987.

Although often loud and conspicuous, Blue Jays are very quiet about their nests. The nest itself, a large, deeply cupped structure, is usually well concealed, often in dense thorn bushes or coniferous thickets. Because of these secretive nesting habits, most Blue Jay breeding records in West Virginia were "confirmed" by presence of fledged young or food for young.

The Blue Jay was the fifteenth most frequently reported species in the Atlas study, and it is well distributed throughout West Virginia.

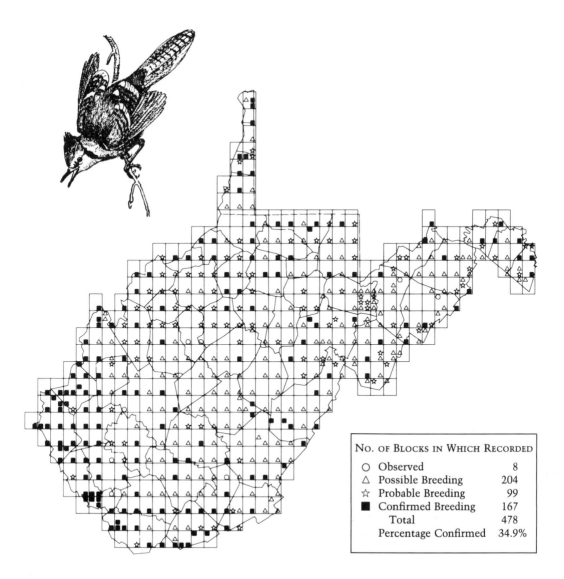

No. of Blocks in Which Recorded	
○ Observed	8
△ Possible Breeding	204
☆ Probable Breeding	99
■ Confirmed Breeding	167
Total	478
Percentage Confirmed	34.9%

American Crow *Corvus brachyrhynchos*

The AMERICAN CROW was the eighth most frequently reported bird in the West Virginia Atlas project. The crow inhabits all of eastern North America, from central Texas to the coniferous forest of northern Ontario and central Quebec. Although it occupies the interior and edges of all kinds of forest, it prefers mixed open and woodland habitat in agricultural areas.

American Crow nests are usually placed in the crotch of a tree near the trunk, 5 to 20 meters from the ground. Almost any kind of tree may serve as a nest tree, but conifers are preferred. The species is very quiet at the nest; incubating birds sit tight and slip off silently when disturbed. As a result, many Atlas reports were "confirmed" by finding fledged young; few nests were found.

Although crows are not as numerous in spruce forest and northern hardwood forest as in the lowlands (Hall 1983), a lack of records for this species in a priority block probably indicates either no Atlas coverage or very light coverage. There was a 1.6 percent (p <0.05) median decrease in American Crows on West Virginia BBS routes over the years 1966 through 1989. The decline may be due to a loss of agricultural land in the state. The eastern United States as a whole showed very slight but significant increase in crow populations on BBS routes (Robbins, Bystrak, and Geissler 1986).

Robbins, Bystrak, and Geissler (1986) show the highest density of crows to be outside of the Allegheny Mountains Region and southwestern parts of the Western Hills Region of West Virginia. The same pattern of distribution can be seen on the Atlas map.

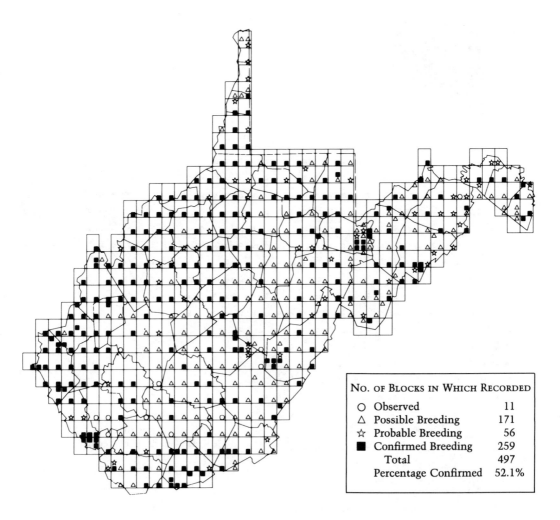

No. of Blocks in Which Recorded	
○ Observed	11
△ Possible Breeding	171
☆ Probable Breeding	56
■ Confirmed Breeding	259
Total	497
Percentage Confirmed	52.1%

Fish Crow *Corvus ossifragus*

A bird of eastern and southern coastal marshes, beaches, and banks of tidal rivers, the FISH CROW has extended its breeding range inland up major rivers in this century. It is found in the upper Hudson River Valley in New York and in central Pennsylvania in the Susquehanna River Valley. The main breeding range is from New York and Massachusetts south along the Atlantic and Gulf coasts to Florida and southern Texas. Robbins, Bystrak, and Geissler (1986) found a significant increase in Fish Crow reports from eastern BBS routes for the years 1965 through 1979.

According to Hall (1983), the Fish Crow occurs in West Virginia only along the Shenandoah River in Jefferson County and along the Potomac River at least as far upstream as Hampshire County. Atlas workers found it farther inland in Grant County on North and South Mill creeks, the headwaters of the South Branch of the Potomac River, and in Mineral and Hampshire counties near tributaries of the North Branch of the Potomac. Most records came from Berkeley and Jefferson counties in the extreme eastern part of the Eastern Panhandle. The only "confirmed" record was of fledged young in Berkeley County.

Fish Crows generally breed in isolated pairs or in small colonies of two or three pairs, usually away from American Crows. The species' nest is similar to that of the American Crow, and many observers fail to distinguish between the Fish Crow and the American Crow. Fish Crow calls also can be confused with the weak calls of young American Crows. For these reasons, it is likely that some Fish Crows were missed or confused with the American Crow. The species cannot be distinguished from size and plumage characteristics unless the two can be closely compared.

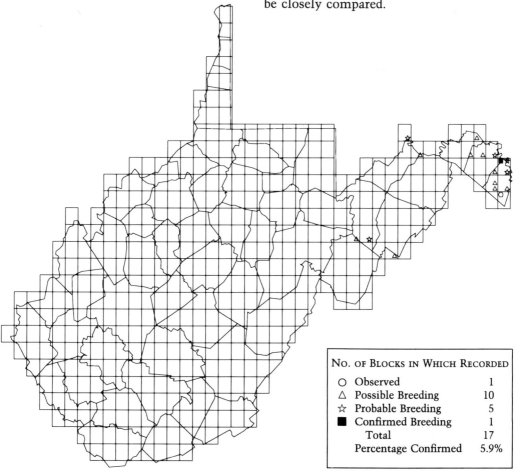

No. of Blocks in Which Recorded	
○ Observed	1
△ Possible Breeding	10
☆ Probable Breeding	5
■ Confirmed Breeding	1
Total	17
Percentage Confirmed	5.9%

Common Raven *Corvus corax*

The COMMON RAVEN has an extensive range over all the continents of the Northern Hemisphere. In eastern North America, ravens are found from northern Canada south through the highlands to northwestern Georgia (AOU 1983). Their large, bulky nests are found on secluded cliffs and, less often, in the tops of tall conifers. Cliff nest sites, typically at least 15 meters above a sheer drop, are usually dark and sheltered by an overhang or vegetation.

The species is increasing in numbers and expanding its range in the Appalachians. Robbins, Bystrak, and Geissler (1986) noted significant increases for the East, and BBS data show an average annual increase of 12 percent ($p < 0.05$) on the ten BBS routes in West Virginia that reported Common Ravens from 1966 to 1987. The population increase may have resulted from an increase in white-tailed deer populations and the resulting greater availability of road-killed carrion. Relief from persecution may also be a factor.

According to Hall (1983), the raven is a regular permanent resident in all the mountain counties. Atlas workers also found breeding evidence throughout the Allegheny Mountains Region, with a few strays to the west in Clay, Harrison, and Monongalia counties. A "possible" record on the Shenandoah River in Jefferson County is just south of the location where in 1981 a pair nested on a cliff on the Potomac near Shepherdstown, at an elevation of less than 120 meters (Hall 1983). The records in Summers and Raleigh counties indicate that the small populations in the gorges of those counties, noted by Hall, are still in existence.

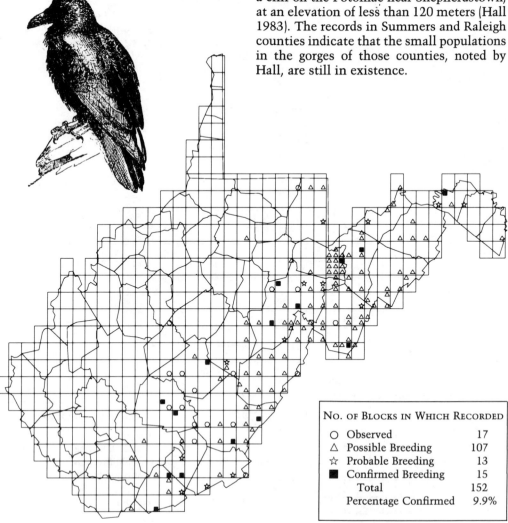

NO. OF BLOCKS IN WHICH RECORDED	
○ Observed	17
△ Possible Breeding	107
☆ Probable Breeding	13
■ Confirmed Breeding	15
Total	152
Percentage Confirmed	9.9%

Black-capped Chickadee *Parus atricapillus*

The BLACK-CAPPED CHICKADEE nests from central and eastern Canada south to central Ohio, southern Pennsylvania, and northern New Jersey, and to eastern Tennessee and western North Carolina in the Appalachian Mountains (AOU 1983). In West Virginia, this chickadee breeds in the Allegheny Mountains Region, in the Ridge and Valley Region through Morgan and Berkeley counties, and in Nicholas, Summers, Mercer, and McDowell counties to the west and south. Atlas workers also found "possible" breeding in the Northern Panhandle near populations in Ohio and Pennsylvania. Hall (1983) reported only a few records for elevations below 300 meters, in valleys within the main range. West of the mountains, Black-caps are not typically found below 580 meters; they are found at somewhat lower elevations to the east.

Black-caps prefer mixed deciduous-coniferous forest and northern hardwoods in West Virginia; they also use residential areas, parks, and small woodlots. They nest in dead standing trees, fence posts, bird boxes, old woodpecker holes, and natural cavities.

The Black-capped Chickadee is often confused with the Carolina Chickadee. Atlas workers found both in 13 counties (see Carolina Chickadee account). In the contact zone, hybrid chickadees sing intermediate songs or both songs (Ward and Ward 1974) and may be intermediate in plumage characteristics (Hall 1983), making identification difficult.

West Virginia BBS data showed an increase in Black-caps of 3.5 percent ($p < 0.01$) per year between 1966 and 1989, a trend that slowed to a 0.5 percent annual increase in the decade of the 1980s. Hall (1983) reported increased winter incursions of Black-capped Chickadees in the late 1970s, with birds remaining to breed at elevations lower than those of previous years. Atlas results likewise indicate the species may have expanded its breeding range in the state in recent years. Atlas workers found the species in both Morgan and Berkeley counties, while Hall (1983) found it to be absent there.

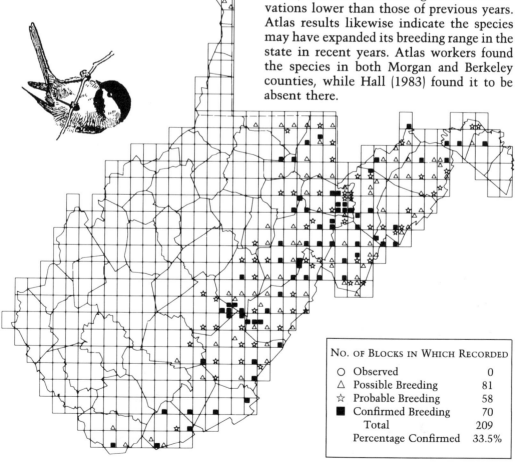

NO. OF BLOCKS IN WHICH RECORDED	
○ Observed	0
△ Possible Breeding	81
☆ Probable Breeding	58
■ Confirmed Breeding	70
Total	209
Percentage Confirmed	33.5%

Carolina Chickadee *Parus carolinensis*

The CAROLINA CHICKADEE is somewhat smaller in size than the Black-capped Chickadee. Its wings and tail are shorter, and it has less white on the margins of its wing and wing coverts than does the Black-cap. Many observers have difficulty distinguishing between them in the field.

Carolina Chickadees breed from central Illinois, central Indiana, central Ohio, southwestern Pennsylvania, and southern New Jersey south to the Gulf Coast and central Florida (AOU 1983). In West Virginia, they are absent from the higher elevations of the Allegheny Mountains and western parts of the Ridge and Valley Region (Hall 1983).

Black-capped and Carolina Chickadees were sympatric, occupying the same Atlas blocks, in 13 counties: Monongalia, Taylor, Upshur, Nicholas, Fayette, Greenbrier, Summers, Monroe, Mercer, and McDowell counties in the west; Hancock County in the Northern Panhandle; and Morgan and Berkeley counties in the Eastern Panhandle. The two species come into contact in Virginia and southwestern Pennsylvania, where they may form hybrids (Johnston 1971; Ward and Ward 1974). Farther west and in the southern Appalachians, the two species are separated by a few kilometers (Brewer 1963) or by some elevation (Tanner 1952). Hall noted that chickadees are scarce in the contact zone on Chestnut Ridge in Monongalia County and on other mountains in West Virginia. No attempt was made by Atlas workers to observe chickadee interactions in the zone of contact. Now that the zone is better defined, the nature of such interaction might be explored.

There is no evidence for population change in Carolina Chickadees in West Virginia over the past few decades, although the species has increased in the eastern states while decreasing in the eastern Ohio hill country (Robbins, Bystrak, and Geissler 1986).

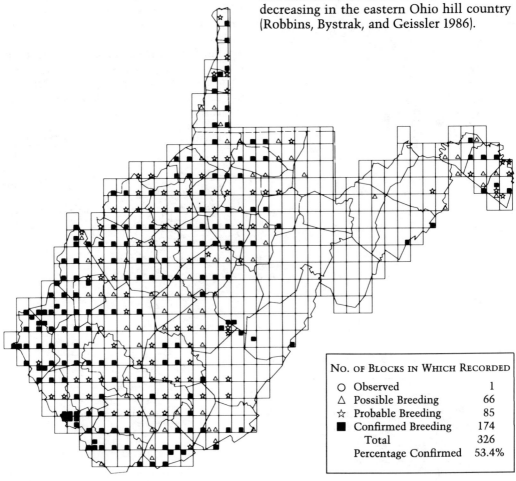

No. of Blocks in Which Recorded	
O Observed	1
△ Possible Breeding	66
☆ Probable Breeding	85
■ Confirmed Breeding	174
Total	326
Percentage Confirmed	53.4%

Tufted Titmouse *Parus bicolor*

The sprightly TUFTED TITMOUSE inhabits rich woodlands and residential areas from Texas, the Gulf states, and Florida north to the Great Lakes, central New York, and central New England (AOU 1983). Although its range is expanding northward, its numbers in Ohio and West Virginia seem to be declining (Robbins, Bystrak, and Geissler 1986). There was an average annual decrease of 1.2 percent ($p < 0.1$) in the West Virginia BBS from 1966 to 1989.

For nesting, Tufted Titmice choose old woodpecker holes, natural cavities, or nest boxes. Their constant activity and distinctive calls make them relatively easy to find. West Virginia Atlas volunteers "confirmed" breeding of Tufted Titmice in every county, but they were less well distributed in the Allegheny Mountains than at lower elevations. They are not as common in spruce forest and mixed spruce-hardwoods as in lowland forest types (Hall 1983).

The Tufted Titmouse was the eleventh most frequently reported species. It tied with the Scarlet Tanager, at 483 blocks occupied.

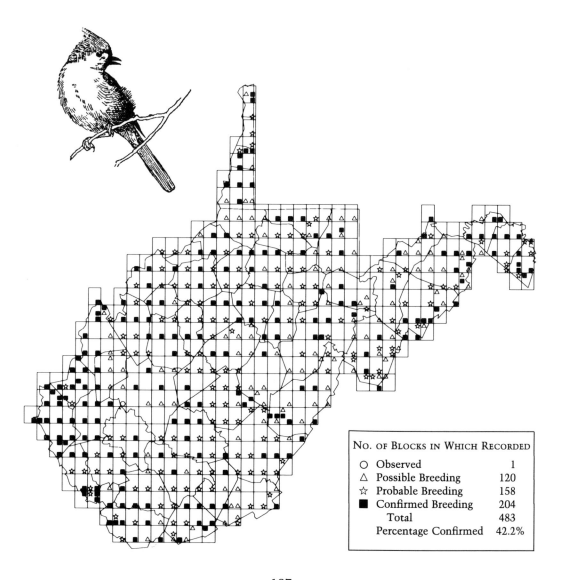

NO. OF BLOCKS IN WHICH RECORDED	
○ Observed	1
△ Possible Breeding	120
☆ Probable Breeding	158
■ Confirmed Breeding	204
Total	483
Percentage Confirmed	42.2%

Red-breasted Nuthatch *Sitta canadensis*

The RED-BREASTED NUTHATCH is much less familiar than its relative the White-breasted Nuthatch. Except during winter, it is a bird of the northern coniferous forest. Its breeding range extends from central Manitoba, northcentral Ontario, central Quebec, and Newfoundland, south to eastern Minnesota, southern Wisconsin, southern Michigan, southern Ontario, northern Pennsylvania, and southern New England, with an extension along the Appalachian Highlands south to North Carolina and Tennessee (AOU 1983). The Breeding Bird Survey showed no significant change in the continental population in the years 1966 through 1978 (Robbins, Bystrak, and Geissler 1986). BBS data for West Virginia showed no significant trend for these years; however, only four routes recorded the species.

West Virginia Atlas workers found the Red-breasted Nuthatch limited to the higher areas of the Allegheny Mountains Region. Records came from Tucker, Grant, Randolph, Pocahontas, Greenbrier, Webster, and Nicholas counties. Hall (1983) also reported summer records from Hampshire and Pendleton counties. In that area, the bird is found in the pure spruce forest, if the trees are tall enough, and in the mixed spruce-hardwoods forest.

The nuthatch nests in natural tree cavities or old woodpecker holes. Atlas workers reported five nests, an excellent number for this species.

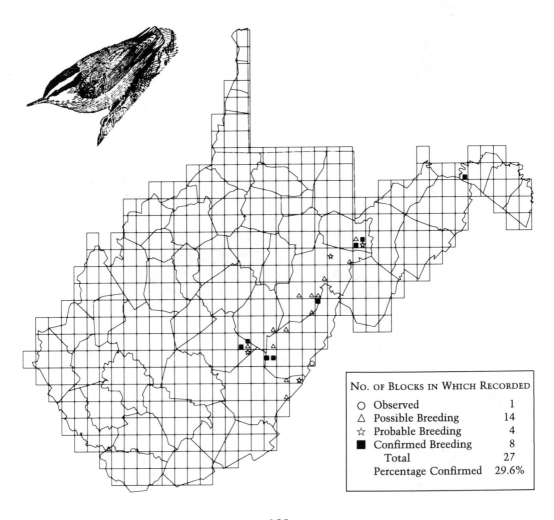

NO. OF BLOCKS IN WHICH RECORDED	
○ Observed	1
△ Possible Breeding	14
☆ Probable Breeding	4
■ Confirmed Breeding	8
Total	27
Percentage Confirmed	29.6%

White-breasted Nuthatch *Sitta carolinensis*

The picture of the WHITE-BREASTED NUT-HATCH creeping head first down a tree trunk is familiar to most people, even nonbirders. The bird itself is also well known to people who operate winter feeding stations, but it is less familiar as a nesting species.

The White-breasted Nuthatch is a widespread but not especially numerous bird of the eastern deciduous forest. Its breeding range extends from northern Minnesota, northern Wisconsin, northern Michigan, southern Ontario, southern Quebec, and Nova Scotia, south to the Gulf Coast and northern Florida (AOU 1983). Eastern populations showed little overall change in the period from 1966 through 1978, but there were local increases and decreases within the region (Robbins, Bystrak, and Geissler 1986).

The Atlas project found the White-breasted Nuthatch throughout the state. There were some unexpected blank spots in the map, but these may be either accidents of limited block coverage or, since the nuthatch is often fairly inconspicuous after mid-June, timing of fieldwork. Some Atlas workers, particularly block busters, may have taken to the field too late to list this species.

The White-breasted Nuthatch is found throughout the deciduous forest region in mature forest and in parks or other more open forest. It usually places its nest in a tree cavity, either natural or an old woodpecker hole, but it will also use nest boxes.

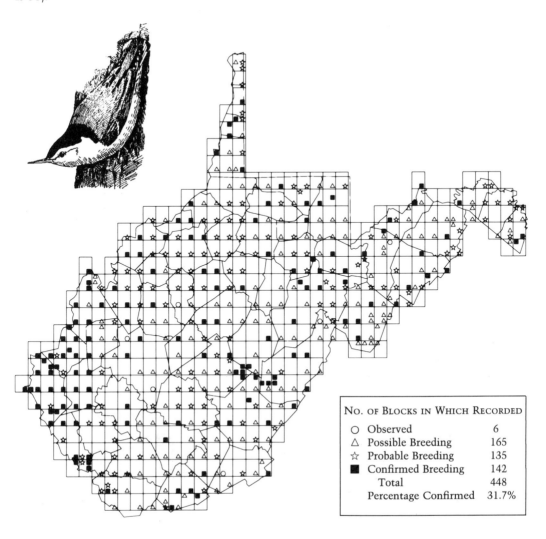

NO. OF BLOCKS IN WHICH RECORDED	
○ Observed	6
△ Possible Breeding	165
☆ Probable Breeding	135
■ Confirmed Breeding	142
Total	448
Percentage Confirmed	31.7%

Brown Creeper *Certhia americana*

The BROWN CREEPER, well concealed by its protective coloration, is easily overlooked as it works its way up a tree trunk. Its high-pitched, low-volume call note is also inaudible to some people.

The Brown Creeper has been considered a northern species, with a breeding range extending from central Manitoba, central Ontario, southern Quebec, and Newfoundland, south to southeastern Missouri, southern Illinois, central Michigan, eastern Ohio, West Virginia, and the lowlands of Virginia and Maryland, with an extension along the Appalachian Highlands to Tennessee and North Carolina (AOU 1983). Occasional nestings occur outside of this general range (Hall 1969). Northeastern populations increased at a rate of 2.7 percent per year (p <0.05) from 1966 to 1978, but there was no significant change from 1978 through 1987 (Robbins, Sauer, Greenberg, and Droege 1989).

Most of the Atlas records were, as expected, in the Allegheny Mountains Region, but there were also a number of "probable" reports from other parts of the state. Several came from the Eastern Panhandle, where the species once nested near the state's lowest elevation (Hall 1969). The Maryland Atlas had records in the adjoining part of the Maryland lowlands (Maryland Atlas data). The reports from the Northern Panhandle are also not unprecedented.

Brown Creepers build their nests under overhanging slabs of bark on dead or dying trees. This species, often thought to be a bird of the spruce forest, may nest in West Virginia at lower elevations, frequently in trees killed by flooding or in elms that have died from Dutch elm disease.

The actual number of breeding blocks may be underrepresented in the Atlas map. This creeper is not conspicuous, and many birders do not recognize its melodious song.

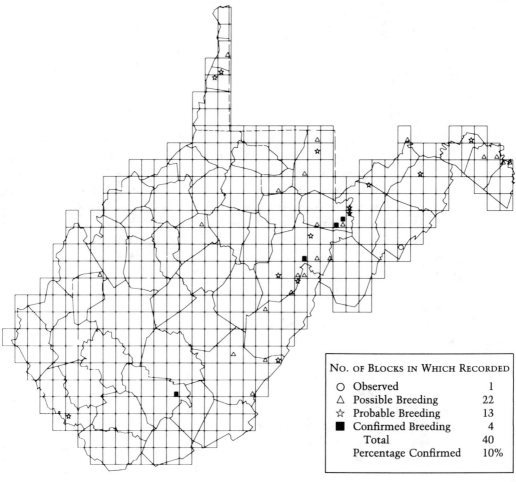

No. of Blocks in Which Recorded	
O Observed	1
△ Possible Breeding	22
☆ Probable Breeding	13
■ Confirmed Breeding	4
Total	40
Percentage Confirmed	10%

Carolina Wren *Thryothorus ludovicianus*

The loud, clarion song of the CAROLINA WREN can be heard at almost any time of the year, as this confiding species winters throughout its range. This wren occurs from southeastern Minnesota, southern Wisconsin, southern Michigan, southern Ontario, southwestern Quebec, central New York, and southern Vermont and Massachusetts, south to the Gulf Coast and southern Florida (AOU 1983). In the northern part of this range, populations oscillate, reflecting heavy mortality during unusually cold and icy winters, as indicated in the BBS data (Robbins, Bystrak, and Geissler 1986). Prolonged, unusual cold alone does not seem to affect the species greatly, but long periods of deep snow cover or icing conditions that trap them in their night roosts in brush piles can cause great havoc. The wren may be totally eliminated from some areas for several years, but after a series of mild winters, populations build back up.

During the Atlas years, the Carolina Wren was found throughout West Virginia. It was missing at high elevations and in some of the mountain valleys. The BBS data showed little overall change in West Virginia populations during the years from 1966 through 1989, but from 1980 to 1989 its numbers increased at a rate of 21 percent per year ($p < 0.1$).

The Carolina Wren nests in open deciduous woodlands, brushy thickets, parks, and residential areas. Its nest is a bulky mass of twigs, fibers, and other plant material placed in a variety of sites, such as old woodpecker holes, crannies in buildings, birdhouses, and upturned roots. The rate of "confirmation" was high because of the wren's habit of frequent nesting near houses.

West Virginia Carolina Wren populations are eliminated only by the coldest winters. The winters of the late 1970s wiped out most of the population in the northern part of the state and at high elevations, and populations were decreased elsewhere. Since then, these wrens have recovered, as shown by the 21 percent increase during the 1980s noted above.

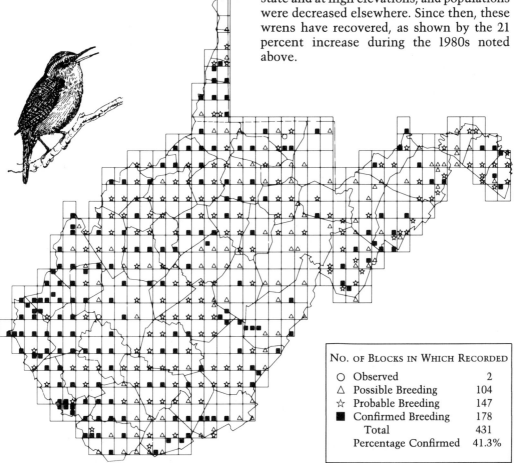

NO. OF BLOCKS IN WHICH RECORDED	
○ Observed	2
△ Possible Breeding	104
☆ Probable Breeding	147
■ Confirmed Breeding	178
Total	431
Percentage Confirmed	41.3%

Bewick's Wren *Thryomanes bewickii*

The BEWICK'S WREN nests in open woodland or brushy thickets, particularly favoring old farm buildings. Its nest is an untidy mass of sticks, chips, and leaves packed into a cranny or crevice (Harrison 1975).

Bewick's Wren was once common through most of its range, but the eastern population has declined so greatly that it is almost gone from large parts of its original range. The Appalachian subspecies, whose type locality is at Philippi in Barbour County, West Virginia, has been listed in the *Federal Register* (1989) as a candidate for addition to the list of endangered forms. The Bewick's Wren has also been listed as a species of special concern in West Virginia (W. Va. DNR n.d.). This wren's range in eastern North America extends from southeastern Minnesota, southern Wisconsin, southern Michigan, southern Ontario, northern Ohio, and eastern Pennsylvania, south to central Alabama, central Georgia, and central South Carolina (AOU 1983). In much of the eastern part of that range, the Bewick's Wren is now very local or extirpated.

Historically, this wren was a fairly common dooryard bird in much of West Virginia, but in the latter part of the nineteenth century it began to decline. At the same time, the House Wren advanced southward into the state. By the 1970s, the Bewick's Wren was numerous only in the dry valleys of the Ridge and Valley Region, with many scattered records elsewhere (Hall 1983).

The Atlas project showed that this situation has changed for the worse. Atlas records for this wren occur in only six blocks, scattered throughout the state, and "confirmed" records appear in only three of these. The species was found on 15 BBS routes from 1966 through 1980, but none has been recorded since then.

Competition with the House Wren is probably not the sole cause for the decline of the Bewick's Wren; the two species seemed to have reached a point of equilibrium by the 1950s. Unless the causes for the decline can be established, little can be done to improve the status of the Bewick's Wren.

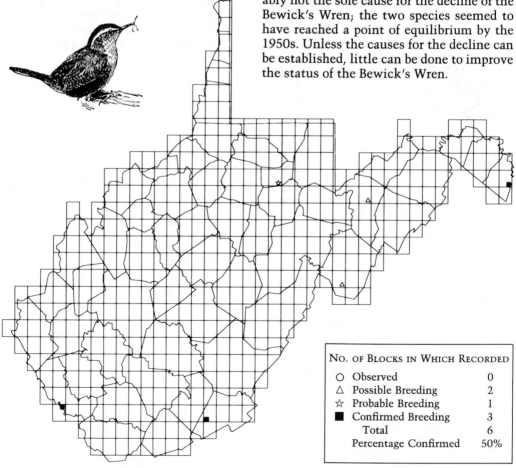

NO. OF BLOCKS IN WHICH RECORDED	
○ Observed	0
△ Possible Breeding	2
☆ Probable Breeding	1
■ Confirmed Breeding	3
Total	6
Percentage Confirmed	50%

House Wren *Troglodytes aedon*

In most of North America, the HOUSE WREN is one of the most familiar dooryard birds. Its English name is derived from its willingness to nest in close proximity to human settlement. This species builds its nest in tree cavities or in bird boxes, and will nest in almost any open or semi-open habitat at all elevations. It is particularly common about farmhouses and suburban homes.

The House Wren nests from southern Manitoba, central Ontario, southwestern Quebec, Maine, and New Brunswick, south to southern Tennessee, northeastern Georgia, northwestern South Carolina, and eastern North Carolina (AOU 1983). Eastern populations showed a slight increase from 1966 through 1978 (Robbins, Bystrak, and Geissler 1986). BBS data indicate that West Virginia populations have been steady since 1966, but have increased at a rate of 6.8 percent per year (*p* <0.01) since 1980.

Although the House Wren historically has been a part of the West Virginia avifauna only since the latter part of the nineteenth century, it was thought to nest in every county by the 1980s (Hall 1983).

Although Atlas workers found the House Wren to be widespread through most of the state, some surprising absences were noted. Away from the Ohio Valley, it was not widespread in the Western Hills. No records were listed in Kanawha, Boone, Mingo, or McDowell counties, and the species occurred in only one block in Jackson, Lincoln, and Logan counties. The absence in Kanawha County must be due, in part, to limited coverage, since Handley (1976) listed it as "fairly common." In general, because of the bird's willingness to nest near humans, the percentage of Atlas "confirmations" is high.

Despite its popularity with most people, there is a dark side to the House Wren: It has a habit of visiting the nests of most birds in its neighborhood and puncturing their eggs. With some species of warblers, this may be an important factor in determining nesting success and local population.

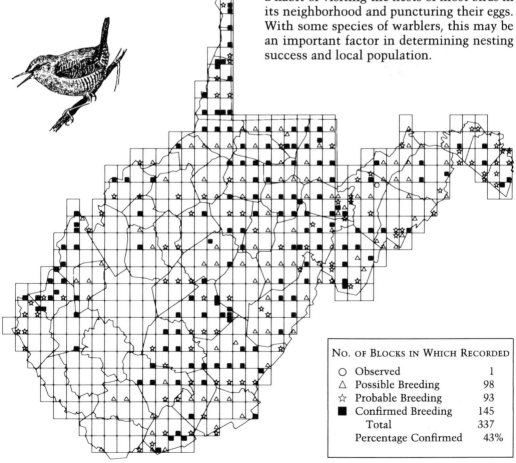

NO. OF BLOCKS IN WHICH RECORDED	
○ Observed	1
△ Possible Breeding	98
☆ Probable Breeding	93
■ Confirmed Breeding	145
Total	337
Percentage Confirmed	43%

Winter Wren *Troglodytes troglodytes*

The tiny WINTER WREN is a characteristic bird of the northern coniferous forest, and its enchanting song is one of the delights of that forest. It nests from central Manitoba, central Ontario, central Quebec, southern Labrador, and Newfoundland, south to central Minnesota, central Wisconsin, central Michigan, southern Ontario, northern Pennsylvania, and northern New Jersey, with an extension through the Appalachian Highlands south to northern Georgia (AOU 1983). Eastern populations decreased at an annual rate of 7.1 percent (p <0.01) from 1966 through 1978, but they increased at an annual rate of 7 percent (p <0.01) between 1978 and 1987 (Robbins, Sauer, Greenberg, and Droege 1989). The decrease was largely a result of mortality in the hard winters of the late 1970s, and the increase was the natural rebound from that loss.

In the past, summer records in West Virginia for the Winter Wren came only from Preston, Tucker, Grant, Pendleton, Randolph, Webster, and Pocahontas counties, with one stray occurrence in Lewis County. There were only two definite nesting records (Hall 1983). Atlas workers recorded the species in all these counties as well as in Nicholas and Fayette. The only "confirmation" was a nest at Kumbrabow State Forest in 1985. Only four BBS routes list Winter Wrens, so the stated annual rate of increase of 3.64 percent, even though statistically significant, may not be biologically meaningful. The most studied population in the state, the one on Shaver's Mountain, has fluctuated, usually tracking the severity of the winter weather. At this writing, the population is once again on the high side.

Atlas workers found the Winter Wren in spruce forest, spruce-hardwoods transition forest, and cooler ravines with hemlock-hardwoods forest. The bird's nest is a ball-like mass of roots, fibers, and stems, and it is concealed in upturned roots of fallen trees, in stump roots, and in mossy hummocks or rocky crevices (Harrison 1975). The lack of "confirmations" is explained by the Winter Wren's concealment of its nest.

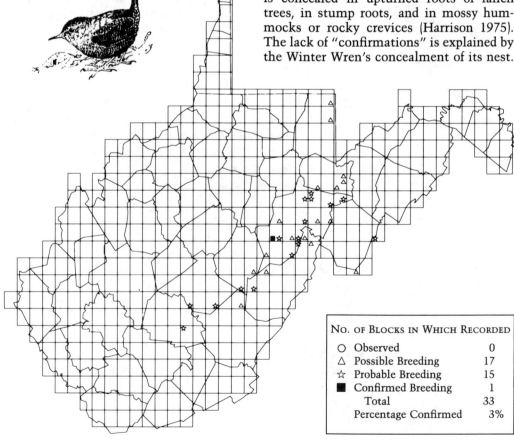

NO. OF BLOCKS IN WHICH RECORDED	
O Observed	0
△ Possible Breeding	17
☆ Probable Breeding	15
■ Confirmed Breeding	1
Total	33
Percentage Confirmed	3%

Sedge Wren *Cistothorus platensis*

The SEDGE WREN is a bird of the wet grass and sedge meadows. It builds its nest, a globular affair made of sedge leaves placed in a dense sedge mass, near the surface of the water.

This species occurs locally from southern Manitoba, western and southern Ontario, northern Michigan, extreme southwestern Quebec, central Maine, and New Brunswick, south to southern Illinois, central Kentucky, westcentral West Virginia and southeastern Virginia (AOU 1983). In the years 1966 through 1978, eastern populations declined (Robbins, Bystrak, and Geissler 1986).

The only previous nesting record for the Sedge Wren in West Virginia was in Kanawha County in 1976 (Shreve 1977). Atlas workers found a nest with young in Barbour County and a "possible" record in Brooke County. In the past, summer records were reported from several other parts of the state where they no longer occur. It is possible that some birds still nest, as they did in the 1950s, in the almost inaccessible wetlands of the northern part of Canaan Valley.

This species is listed in *Vertebrate Species of Concern in West Virginia* (W. Va. DNR n.d.) and, as with the Marsh Wren, it is dependent on the continued existence of wetlands.

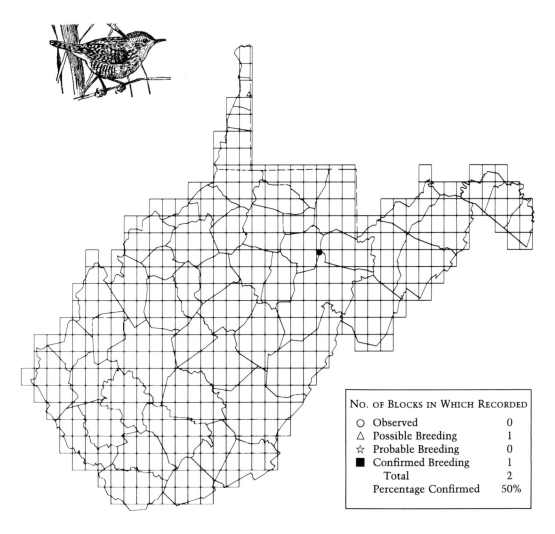

NO. OF BLOCKS IN WHICH RECORDED	
O Observed	0
△ Possible Breeding	1
☆ Probable Breeding	0
■ Confirmed Breeding	1
Total	2
Percentage Confirmed	50%

Golden-crowned Kinglet *Regulus satrapa*

The GOLDEN-CROWNED KINGLET is a species of the northern coniferous forest, nesting from northern Manitoba, northern Ontario, central Quebec, and Newfoundland, south to northern Minnesota, northern Michigan, southern Ontario, central New York, northern Pennsylvania, southern New England, and along the higher elevations of the Appalachians south to Tennessee and North Carolina (AOU 1983). It builds its nest in mature coniferous forest, particularly spruce, and in recent years has been expanding its range southward to occupy spruce plantations that have matured enough for the species. The Pennsylvania Atlas project found it more widespread through the northern and mountainous part of that state than expected (Brauning 1992).

Although overall population trends are not significant, the Golden-crowned Kinglet underwent a sharp decline in the late 1970s as a result of extraordinarily cold weather on its wintering grounds in the southeastern United States. Populations seem to have recovered and may now be increasing.

Atlas workers found the Golden-crowned Kinglet nesting only in the higher parts of the Allegheny Mountains Region. "Probable" and "confirmed" records came from only Preston, Tucker, Pendleton, Randolph, and Pocahontas counties, all from areas of spruce forest. The incursion of kinglets into spruce plantations has not yet been detected in West Virginia, though some "possible" records may reflect such inroads. Only two BBS routes have reported kinglets, and no population trends can be given. The population in the virgin spruce tract on Shavers Mountain underwent a decline in the late 1970s but has since recovered. In the mid-1980s, populations on Spruce Mountain were high.

Golden-crowned Kinglet nests are small hanging structures fixed to the ends of spruce branches where the foliage is the thickest. They are extremely hard to find, and the two "confirmed" records were of adults carrying food and feeding fledged young.

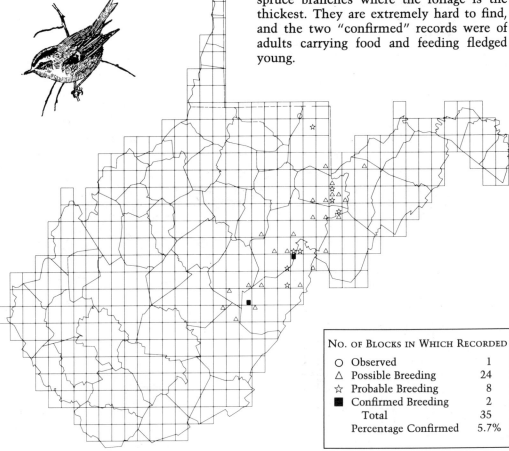

NO. OF BLOCKS IN WHICH RECORDED	
○ Observed	1
△ Possible Breeding	24
☆ Probable Breeding	8
■ Confirmed Breeding	2
Total	35
Percentage Confirmed	5.7%

Blue-gray Gnatcatcher *Polioptila caerulea*

The tiny BLUE-GRAY GNATCATCHER with its high-pitched, inconspicuous song is often overlooked in its habitat in the leafy canopy. This southern species has been expanding its range northward, and it now occurs from southeastern Minnesota, southern Wisconsin, southern Michigan, extreme southern Ontario, central New York, southern Vermont, New Hampshire, and southern Maine, south to the Gulf Coast and southern Florida (AOU 1983). It inhabits the mature stages of a variety of deciduous forest types. Between 1966 and 1978, eastern populations had shown significant increases (Robbins, Bystrak, and Geissler 1986), and these continued through 1987 (Robbins, Sauer, Greenberg, and Droege 1989). For West Virginia, the BBS data show an increase of 2.9 percent annually (*p* <.1) for the period from 1980 through 1989.

The West Virginia Atlas project found the Blue-gray Gnatcatcher in all parts of the state. The center of occurrence was in the Ohio Valley, extending into other parts of the Western Hills. It was found only in scattered blocks through the Allegheny Mountains Region. The bird is most common in the oak-hickory-pine forests of the Ridge and Valley Region, and nesting has been observed at elevations up to about 1,000 meters (Hall 1983).

The nest is a delicate compact cup made of plant fiber held together with insect silk or spiderweb, and the whole is decorated with lichens. It is saddled on a horizontal branch, usually quite high. Nests are not easily found, except during the spring building period, and most Atlas "confirmations" were of feeding young out of the nest, a time when the birds are especially obvious.

Although this species is a Neotropical migrant, many of them winter in the southern United States, and it appears to be in no present danger.

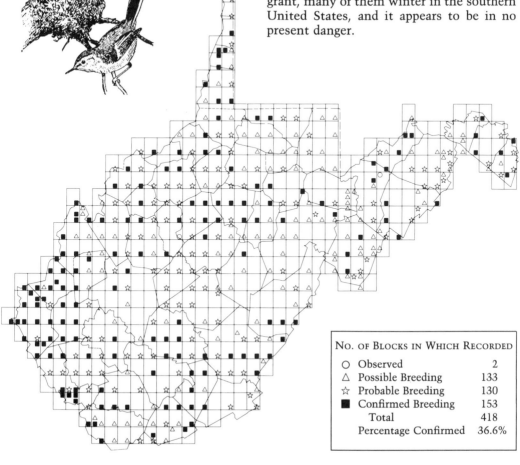

NO. OF BLOCKS IN WHICH RECORDED	
○ Observed	2
△ Possible Breeding	133
☆ Probable Breeding	130
■ Confirmed Breeding	153
Total	418
Percentage Confirmed	36.6%

Eastern Bluebird *Sialia sialis*

In the past, the EASTERN BLUEBIRD was a common and familiar bird to most people throughout eastern North America south of southern Manitoba, central Ontario, and southern Quebec. In recent years, however, populations have declined throughout the country. The introduced European Starling and House Sparrow are usually able to take over nesting sites from bluebirds. On several occasions, particularly in 1957–58, 1976–77, and 1977–78, extremely cold weather caused massive mortality in the southern wintering grounds. In very recent times, the cutting of dead trees for firewood has eliminated many nesting sites, as has the change to metal fence posts. The establishment of bluebird trails—long series of artificial nest boxes—has improved the species' overall situation somewhat. The BBS data (Robbins, Sauer, Greenberg, and Droege 1989) confirm that populations underwent a significant decline between 1966 and 1978, but they have made a good recovery since 1978, increasing 9.8 percent annually ($p < .01$).

Atlas workers in West Virginia found the bluebird to be widespread, nesting throughout the state, even at the highest elevations. This species breeds in open country with occasional trees that supply nesting sites, although it also often uses holes in old fence posts. As indicated above, bluebirds commonly use boxes that are suitably placed with open space on all sides. Because of the Eastern Bluebird's nest locations, the percentage of "confirmed" records is quite high.

West Virginia populations declined at an annual rate of 2.8 percent ($p < .05$) in the years 1966 to 1987 (BBS data), but the trend in the 1980s has been upward, as populations recovered from winter mortality in the late 1970s. In spite of this encouraging trend, the future of the Eastern Bluebird is not assured. Populations and nesting success should be carefully monitored in the future.

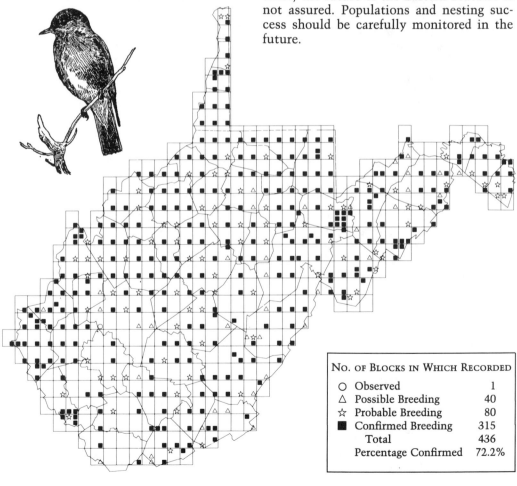

No. of Blocks in Which Recorded	
○ Observed	1
△ Possible Breeding	40
☆ Probable Breeding	80
■ Confirmed Breeding	315
Total	436
Percentage Confirmed	72.2%

Veery *Catharus fuscescens*

The ringing song of the VEERY is the characteristic thrush sound of the transition zone between coniferous and deciduous forest. In eastern North America, Veeries nest from southern Manitoba, southern Ontario, southern Quebec, New Brunswick, and southern Newfoundland, south to Iowa, northern Illinois, northcentral Indiana, northern Ohio, southern Pennsylvania, and south along the Appalachian highlands to North Carolina and Tennessee (AOU 1983). After increasing at an annual rate of 1.6 percent (p <.05) between 1966 and 1978, the eastern populations began to decline at a rate of 2.4 percent annually (p <.05) (Robbins, Sauer, Greenberg, and Droege 1989). Despite the continental decline, West Virginia populations showed a striking 8.8 percent annual increase (p <.05) over the period spanning 1966 through 1989 (BBS data).

For the most part, the West Virginia Atlas project found this species in the Allegheny Mountains Region, where it occurred at lower elevations and was found in more blocks than the Swainson's and Hermit thrushes. A few scattered records were also reported from the northern hardwood forest in the higher, southern part of the Western Hills Region. Veeries occur in mixed spruce-northern hardwoods forest, in hemlock-hardwoods areas, and in northern hardwoods without any conifers. They are most common in the mid-aged, second-growth stages. Hall (1983) described the fluid zones of tension existing between Veeries and the other two northern thrushes that occur in the higher West Virginia mountains.

This species would appear to be under no local threat other than the usual fluctuations in population, but it is vulnerable to loss of winter habitat in the tropics.

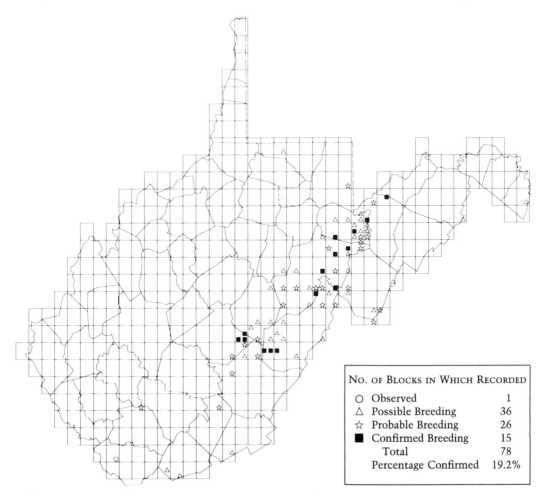

NO. OF BLOCKS IN WHICH RECORDED	
O Observed	1
△ Possible Breeding	36
☆ Probable Breeding	26
■ Confirmed Breeding	15
Total	78
Percentage Confirmed	19.2%

Swainson's Thrush *Catharus ustulatus*

The SWAINSON'S THRUSH is a species of the northern coniferous forest. In eastern North America it ranges from central Manitoba, northern Ontario, northern Quebec, and southern Labrador, south to northern Minnesota, northern Wisconsin, northern Michigan, southern Ontario, southern New York, extreme northeastern Pennsylvania, and northern New England, with disjunct populations in West Virginia and, in recent years, on Mount Rogers in Virginia (AOU 1983).

The Swainson's Thrush is perhaps the most numerous of the spotted thrushes of the genus *Catharus*, but reports of both spring and fall migrations in recent years have indicated a population decline, as have banding records. The BBS data, however, do not show this. From 1966 through 1978, the eastern populations increased at an annual rate of 3.4 percent (*p* <.01). From 1978 to 1987 there was a slight decrease that, although not statistically significant, might indicate the beginning of a decline (Robbins, Sauer, Greenberg, and Droege 1989). Some populations in the far West have undergone declines (Marshall 1988).

For many years, a small but healthy disjunct population has existed in the higher mountains in the spruce belt of West Virginia. In the past, there have been summer records from Tucker, Grant, Pendleton, Randolph, Pocahontas, and Webster counties (Hall 1983). The Atlas project did not record this species in Tucker or Grant counties but did find it in Greenbrier County. "Confirmed" or "probable" records came from Pocahontas, Pendleton, and Randolph counties.

The Swainson's Thrush occurs in the greatest numbers in pure spruce forest of any age, but it is also found in the spruce-northern hardwoods forest. Only one West Virginia BBS route has listed this species, so no quantitative trends for the state are possible. Qualitative observations in the past few years indicate that the population is still healthy (Hall pers. obs.).

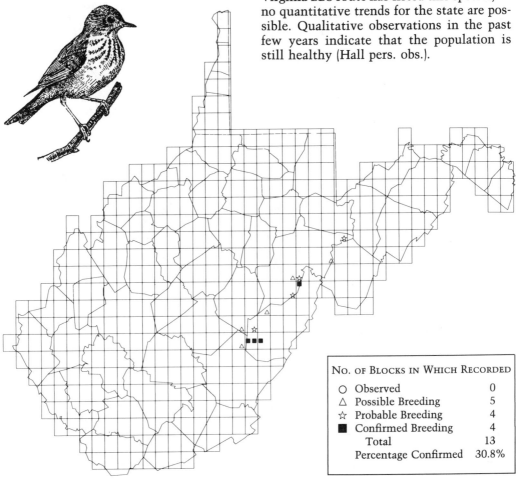

NO. OF BLOCKS IN WHICH RECORDED	
○ Observed	0
△ Possible Breeding	5
☆ Probable Breeding	4
■ Confirmed Breeding	4
Total	13
Percentage Confirmed	30.8%

Hermit Thrush *Catharus guttatus*

The HERMIT THRUSH is often considered the finest singer of the eastern forests. It builds its nest on the ground, making a compact but bulky structure from twigs, ferns, and mosses lined with conifer needles.

This species occurs in the northern forests from northern Manitoba, northern Ontario, northern Quebec, southern Labrador south to northern Minnesota, central Wisconsin, central Michigan, southern Ontario, northeastern Ohio, central Pennsylvania, and southern New England, with a southward extension along the crest of the Appalachians to West Virginia and Mount Rogers, Virginia (AOU 1983). Since 1983, the Hermit Thrush has appeared as far south as eastern North Carolina in the summer. It inhabits coniferous forests and the mixed coniferous-hardwoods forest. It is not limited to mature forests and occurs widely in second growth. Eastern populations showed no significant trends in the period from 1966 through 1978 (Robbins, Bystrak, and Geissler 1986), but decreases resulting from the severe winters of the late 1970s were evident in those years.

Hall (1983) reported nesting records for Preston, Tucker, Grant, Pendleton, Randolph, Pocahontas, and Greenbrier counties. Atlas workers did not find Hermit Thrushes in Preston but did find some in Nicholas County. The absence in Preston is of interest because the Maryland Atlas found the species in several blocks in adjacent Garrett County, and the Pennsylvania Atlas found them right up to the state line in Somerset and Fayette counties (Brauning 1992). In West Virginia, Hermit Thrushes were limited to the higher ridges containing at least some spruce. The species is most numerous above 1,200 meters (Hall 1983).

West Virginia populations fluctuate, possibly because of severe weather on the bird's wintering grounds in the southern United States, but too few BBS routes have reported them to make any quantitative statements. Presumably this species is in no danger in West Virginia as long as some spruce forest exists in the state.

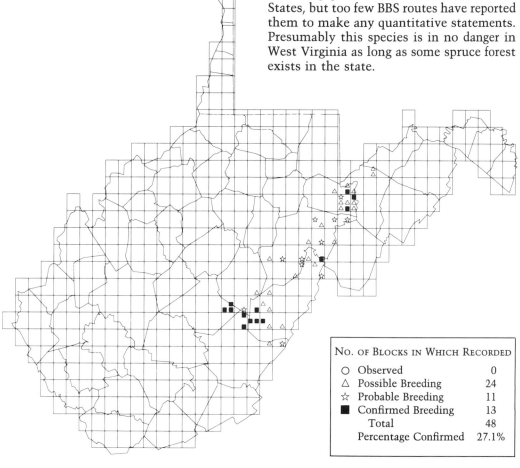

NO. OF BLOCKS IN WHICH RECORDED	
○ Observed	0
△ Possible Breeding	24
☆ Probable Breeding	11
■ Confirmed Breeding	13
Total	48
Percentage Confirmed	27.1%

Wood Thrush *Hylocichla mustelina*

The flutelike voice of the WOOD THRUSH is the evensong of the eastern deciduous forest. The Wood Thrush occurs in mature or near-mature deciduous forests from northern Minnesota, northern Michigan, southern Ontario, and northern New England, south to the Gulf Coast and northern Florida (AOU 1983). From 1966 to 1978, eastern populations increased at an annual rate of 1.3 percent (p <.01), but from 1978 to 1987 they decreased at an annual rate of 4.0 percent (p <.01) (Robbins, Sauer, Greenberg, and Droege 1989). The Wood Thrush seems able to use small forest fragments (Robbins, Dawson, and Dowell 1989). Under such conditions in the Midwest, it is very susceptible to cowbird parasitism (Robinson, 1992), but this may not be true in the East at the present time.

The West Virginia Atlas project found the Wood Thrush to be present in all parts of the state, and the percentage of "confirmations" was high. It occurs in all forest types except the pure spruce, but it does occasionally nest in the spruce-northern hardwoods forest at elevations up to 1,000 meters. The BBS data show no significant trend for West Virginia populations at present.

Thrush nests are compact cups made of stems, moss, and mud, lined with rootlets and placed in a fork or horizontal branch 2 to 15 meters above ground. Approximately one-third of the "confirmations" came from finding nests.

Although the population in the state seems to be stable or even increasing at present, the Wood Thrush is declining in most regions to the north, and its status should be monitored in the future.

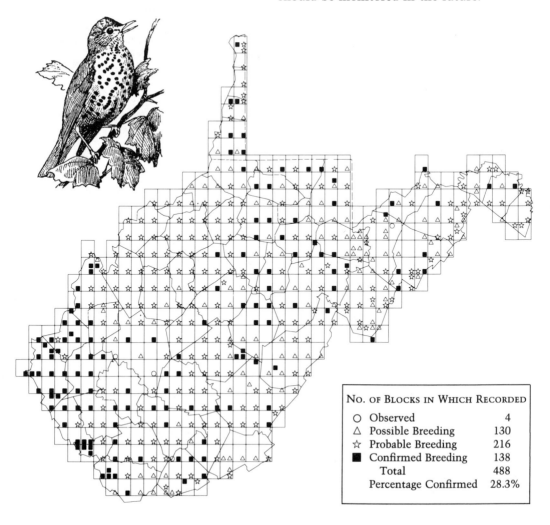

No. of Blocks in Which Recorded	
O Observed	4
△ Possible Breeding	130
☆ Probable Breeding	216
■ Confirmed Breeding	138
Total	488
Percentage Confirmed	28.3%

American Robin *Turdus migratorius*

The AMERICAN ROBIN is without doubt the most familiar bird in the eastern United States. It nests in practically all habitats except the deepest forest interiors, and it occurs throughout the East except in southern Florida. In the period between 1957 and 1983, the southern boundary of the robin's range was extended markedly (AOU 1957; 1983). From 1966 to 1978, populations in the eastern United States increased slightly (Robbins, Bystrak, and Geissler 1986).

It comes as no surprise that Atlas workers found the robin nesting throughout West Virginia. It tied with the Rufous-sided Towhee as the most frequently reported species. The robin nests in forest edge, open woodlands, agricultural areas, house gardens, and even in urban areas that contain trees. Its nest is easy to spot, and the newly fledged young even easier to see, so an extremely high percentage of records (83.5%) were "confirmed." From 1978 to 1989, West Virginia populations increased at 1.3 percent annually (*p* <.10) (BBS data).

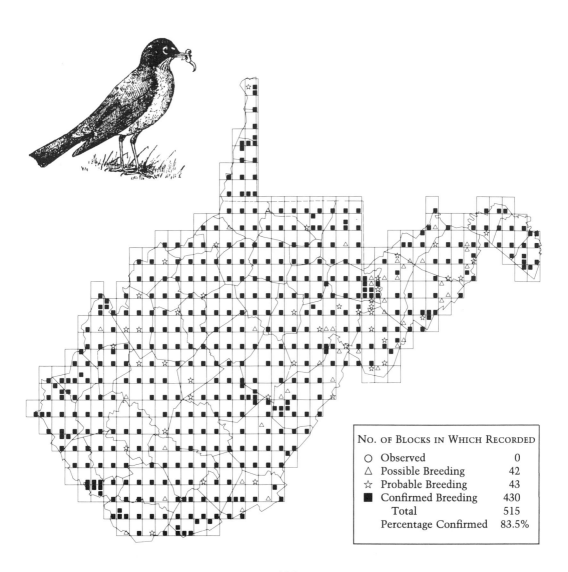

No. of Blocks in Which Recorded	
○ Observed	0
△ Possible Breeding	42
☆ Probable Breeding	43
■ Confirmed Breeding	430
Total	515
Percentage Confirmed	83.5%

Gray Catbird *Dumetella carolinensis*

The GRAY CATBIRD, with its felinelike call note, is familiar to most people in West Virginia. This bird often nests in ornamental bushes in dooryards.

The species breeds from southern Ontario, Quebec, and New Brunswick, south to southern Georgia, southern Alabama, and central Louisiana (AOU 1983). From 1966 to 1978, eastern populations fluctuated in response to cold weather in the wintering area in southeastern United States, but there was no overall significant trend (Robbins, Bystrak, and Geissler 1986). From 1978 to 1987, populations decreased at the rate of 1.4 percent per year ($p < .05$) (Robbins, Sauer, Greenberg, and Droege 1989).

The Atlas map shows the catbird to be widespread in West Virginia, where it occurs in every part of the state. As a result of its confiding habits, it ranks as the seventh most frequently reported species. West Virginia populations had remained stable over the BBS period from 1966 to 1978 (Robbins, Bystrak, and Geissler 1986), but they have increased 3.3 percent per year ($p < .01$) during the 1980s (BBS data).

The catbird nests in dense, brushy situations at forest edges, road cuts, and power lines, as well as in hedgerows and gardens. Its nest, a loose assembly of twigs lined with leaves and grasses, is placed a few feet off the ground in a dense thicket or briar patch.

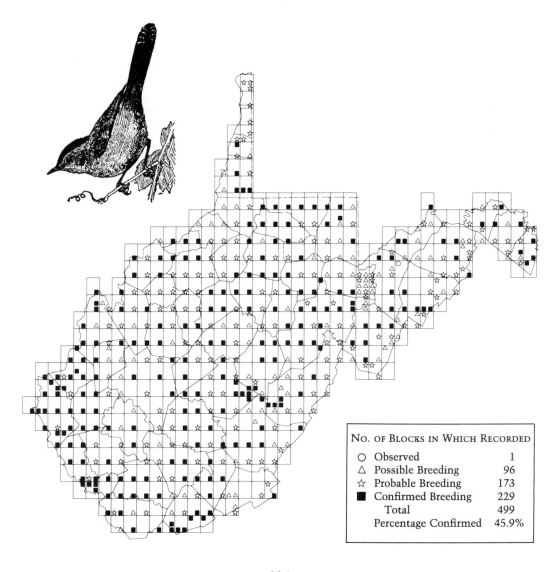

No. of Blocks in Which Recorded	
○ Observed	1
△ Possible Breeding	96
☆ Probable Breeding	173
■ Confirmed Breeding	229
Total	499
Percentage Confirmed	45.9%

Northern Mockingbird *Mimus polyglottos*

The NORTHERN MOCKINGBIRD is a southern species that has been extending its range northward for many years. Its present range includes all of the United States north to southern Iowa, central Illinois, central Indiana, northern Ohio, southern Pennsylvania, southern New York, and New England (AOU 1983), but there are many scattered nesting records north of this. Between 1966 and 1978, eastern populations declined at an annual rate of 1.7 percent ($p < .01$), but from 1978 to 1987 populations remained unchanged (Robbins, Sauer, Greenberg, and Droege 1989).

West Virginia Atlas workers found the mockingbird to be most frequent in the Ohio Valley south of Wetzel County, and in the Shenandoah and Potomac valleys. It occurred in the Northern Panhandle, but nesting was not "confirmed." The species was found at only a few locations in the valleys in the Allegheny Mountains and in the high southern part of the Western Hills. Throughout the northern part of the Western Hills, it was widespread but not universal. BBS data show little change over the period from 1966 through 1987 for West Virginia populations, but in the 1980s there was a 12 percent annual increase ($p < .05$).

Mockingbirds are a conspicuous member of the local avifauna that nest in open or partly open areas, with scattered trees and light brushy patches. Fields infested with multiflora rose, which supply both cover and winter food, are favorite locations. The species' tendency to nest near houses resulted in a high percentage of "confirmations." Unless the loss of brushy areas becomes a major factor, this species should increase in numbers northward in the future.

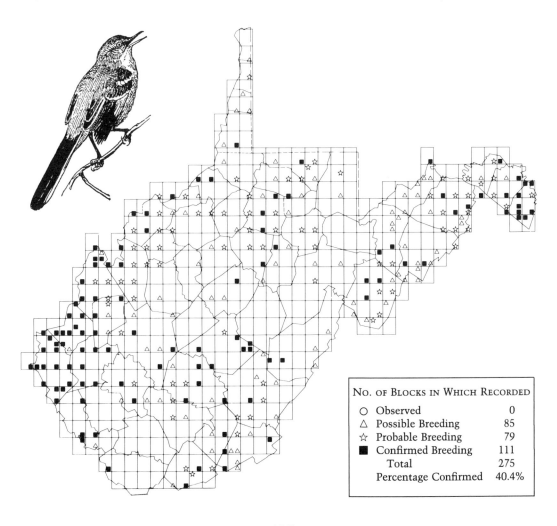

No. OF BLOCKS IN WHICH RECORDED	
○ Observed	0
△ Possible Breeding	85
☆ Probable Breeding	79
■ Confirmed Breeding	111
Total	275
Percentage Confirmed	40.4%

Brown Thrasher *Toxostoma rufum*

The Brown Thrasher is a common bird of the eastern United States, but its secretive habits make it unfamiliar to many people. Except upon arrival in the spring, when the males sing incessantly from the tops of trees, these birds spend most of their time in dense brushy fields or roadside brush rows.

The Brown Thrasher's breeding range is throughout eastern North America, south of southern Ontario and Quebec. From 1966 to 1978, eastern populations remained steady, but some declines were noted in a few states, including West Virginia (Robbins, Bystrak, and Geissler 1986).

Atlas workers found the Brown Thrasher to be widespread, and they confirmed the speculation (Hall 1983) that it nested in every county. According to Brooks (1944), Sutton had found it uncommon in the Northern Panhandle, but Buckelew (1976) listed it as fairly common, and the Atlas project found it in every block in that area. It was not located very widely in Kanawha County, but this may be an artifact of coverage. The species is very inconspicuous after nesting starts, and it may therefore be underrepresented.

The Brown Thrasher is an inhabitant of any dense brushy area at all elevations in the state. Its nest is a loose assemblage of twigs placed on the ground at the base of a tree or few feet off the ground (Harrison 1975).

BBS data indicate that from 1966 through 1989 Brown Thrasher populations in West Virginia declined at an annual rate of 4.4 percent ($p < .01$). The species is in some danger if the amount of brushy habitat available in the state decreases.

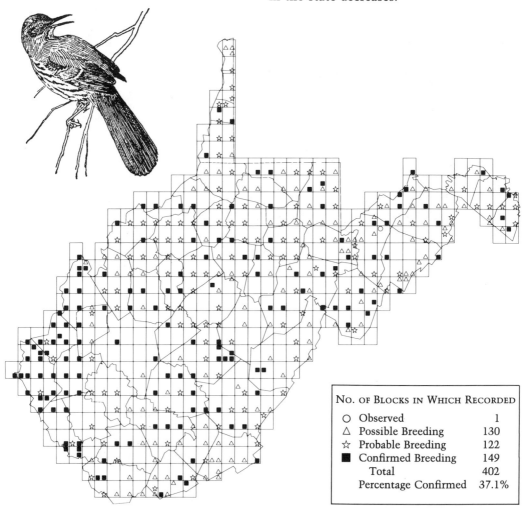

NO. OF BLOCKS IN WHICH RECORDED	
○ Observed	1
△ Possible Breeding	130
☆ Probable Breeding	122
■ Confirmed Breeding	149
Total	402
Percentage Confirmed	37.1%

Cedar Waxwing *Bombycilla cedrorum*

Flocks of CEDAR WAXWINGS are a familiar sight in most parts of West Virginia, but this species does not follow the typical lifestyle of most birds. It may be present in a given region throughout the year or may disappear for varying times at any season of the year. It may not nest in the same places this year that it did last year or will next year. Some waxwings may migrate considerable distances, as shown by a bird banded at Morgantown and recovered in Oaxaca, Mexico. The Cedar Waxwing's general range is from the limit of trees in the North, south to southern Illinois, eastern Tennessee, northern Alabama, and northern Georgia, and to northwestern South Carolina, where it wanders nomadically (AOU 1983).

Waxwings may occur in almost any wooded habitat, but they generally nest in large trees in edge situations or in gardens and farm lots. On occasion, they are almost colonial, with several nests in close proximity. Their nests are loosely woven from twigs, grasses, and weed stems and are placed on a horizontal branch 2 to 6 meters above the ground (Harrison 1975).

Eastern populations of Cedar Waxwings increased slightly from 1966 to 1978 (Robbins, Bystrak, and Geissler 1986). From 1966 to 1989, the West Virginia BBS data showed no significant change.

Atlas workers found waxwings nesting throughout the state; however, the nature of the Atlas methodology makes it impossible to say they were in all areas every summer. Waxwings often nest late in the summer, a period in which many Atlas volunteers had ceased work; the number of "confirmed" records may therefore underrepresent the actual incidence of breeding. Conversely, many of the "possible" records obtained in early June may not represent birds that bred in a given block at a later date.

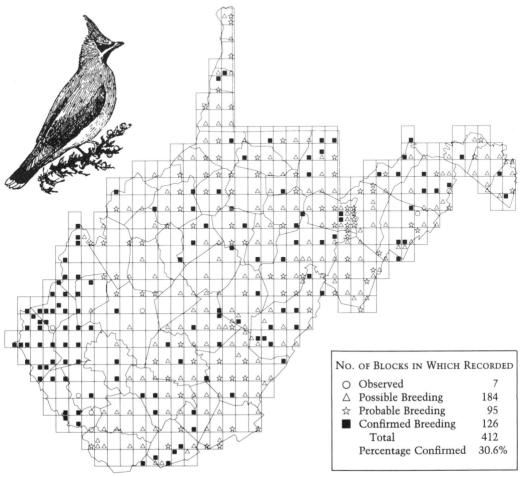

NO. OF BLOCKS IN WHICH RECORDED	
○ Observed	7
△ Possible Breeding	184
☆ Probable Breeding	95
■ Confirmed Breeding	126
Total	412
Percentage Confirmed	30.6%

Loggerhead Shrike *Lanius ludovicianus*

The LOGGERHEAD SHRIKE was never a common bird in the southeastern United States, but populations have declined greatly since the BBS was initiated in 1966. The northern subspecies has been listed in the *Federal Register* (1989) as a candidate for inclusion on the endangered list.

Hall (1983) cited nesting records for eight counties (Berkeley, Hampshire, Hardy, Grant, Pendleton, Greenbrier, Monroe, and Barbour) and summer records for five additional ones. The Atlas workers "confirmed" nesting only in Berkeley, Grant, Greenbrier, Monroe, and Mercer counties, with "probable" reports in Jefferson County. The stronghold of the species seems to be the Great Appalachian Valley (Shenandoah Valley in West Virginia and Virginia). Most West Virginia and Virginia (VSO 1989) records are from that area, as are many of the few Pennsylvania records (Brauning 1992) and many (non-Atlas) reports from Tennessee.

The shrike is a bird of open pasture land, requiring scattered trees for perching and nesting. The habitat in the Shenandoah and Upper Greenbrier valleys is very suitable. The population decline is probably the result of several factors, such as pesticide effects on species that are high in the food chain; changing agricultural practices, which have resulted in the loss of both breeding and wintering habitat; and predation on wintering birds.

The Loggerhead Shrike is included in *Vertebrate Species of Concern in West Virginia* (W. Va. DNR n.d.). However, there seems to be little that can be done to improve the status of this species.

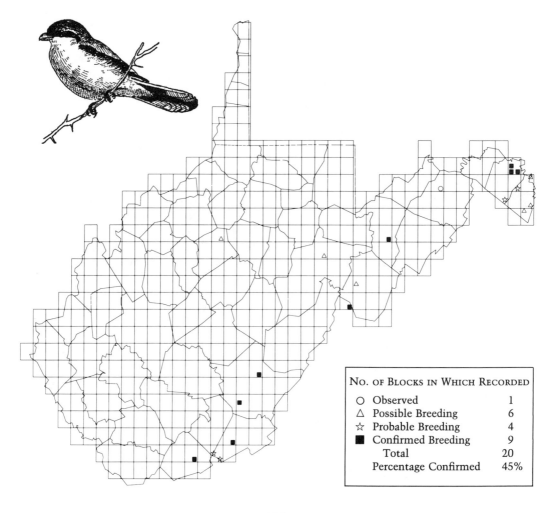

NO. OF BLOCKS IN WHICH RECORDED	
O Observed	1
△ Possible Breeding	6
☆ Probable Breeding	4
■ Confirmed Breeding	9
Total	20
Percentage Confirmed	45%

European Starling *Sturnus vulgaris*

The EUROPEAN STARLING was introduced into North America by the liberation of a few birds in New York City in 1890. Since that time, the species has increased and extended its range to cover nearly the whole of North America south of the boreal tree line. BBS data show that from 1966 to 1978 populations experienced some increases and decreases but generally declined. The occasional declines were attributed to the hard winters of the mid-1970s (Robbins, Bystrak, and Geissler 1986).

The first starlings were seen in West Virginia in 1914 (Hall 1983), and they are now found throughout the state. Nests are often built in buildings and other structures, as well as in tree cavities. Such nests are easily found, and the percentage of "confirmations" is high. Starlings remain scarce at high elevations and in heavily forested areas, but in open farmlands and near towns, they are sometimes abundant. The BBS data show that in West Virginia the species has been declining at an annual rate of 1.73 percent ($p < .10$) during the period from 1966 through 1989.

Most birders tend to ignore the European Starling as a non-native intruder, but it is now an important component of the state's avifauna. Starlings exert a negative influence on the populations of many cavity-nesting birds by ousting them from their nest cavities. The Red-headed Woodpecker population has been particularly vulnerable. Starlings also compete with other species for food resources, but on the positive side they are one of the few predators on the larvae of the Japanese beetle. Large winter roosts or spring and fall aggregations of European Starlings often constitute an esthetic nuisance and, in some cases, may produce health hazards.

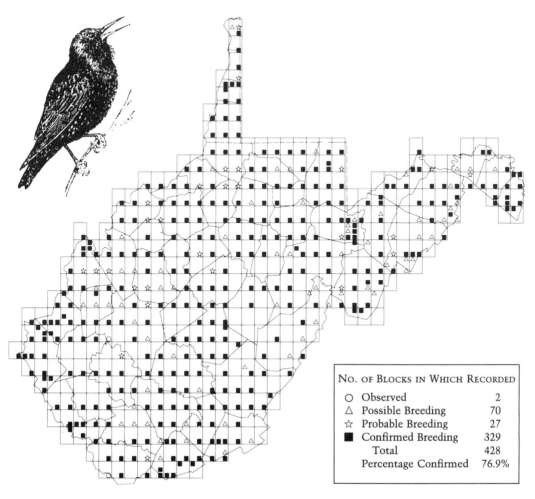

NO. OF BLOCKS IN WHICH RECORDED	
○ Observed	2
△ Possible Breeding	70
☆ Probable Breeding	27
■ Confirmed Breeding	329
Total	428
Percentage Confirmed	76.9%

White-eyed Vireo *Vireo griseus*

The WHITE-EYED VIREO is a southern species that has been extending its range northward for some time. The American Ornithologists' Union (1983) reports that its
range extends from southern Wisconsin,
southern Michigan, southern Ontario,
southern New York, and southern New England, south to the Gulf Coast and southern
Florida. At present the species is advancing
northward from this range. After remaining
constant through the earlier period of the
BBS, the populations have declined at a rate
of 1.2 percent per year ($p < .05$) since 1978
(Robbins, Sauer, Greenberg, and Droege
1989). West Virginia populations have increased during the BBS period and from 1980
through 1989 showed an annual increase of
5.2 percent ($p < .01$) (BBS data).

The Atlas map shows the White-eyed
Vireo to be widespread in the region west of
the mountains. Until recently it had not
been very numerous in the Northern Panhandle, but Atlas workers found it in all
blocks covered there. It is missing at high
elevations, but it occurs locally in the valleys
of the Allegheny Mountains and Ridge and
Valley regions.

Unlike the other vireos, the White-eyed
Vireo inhabits the brushy second-growth
woodlands throughout the southern forest
region. The nest is a cup suspended from a
fork near the end of the branch in a low
shrub, 0.5 to 2 meters off the ground (Harrison 1975). The increased use of herbicide
spray on power-line cuts and roadsides will
have a negative effect on this species.

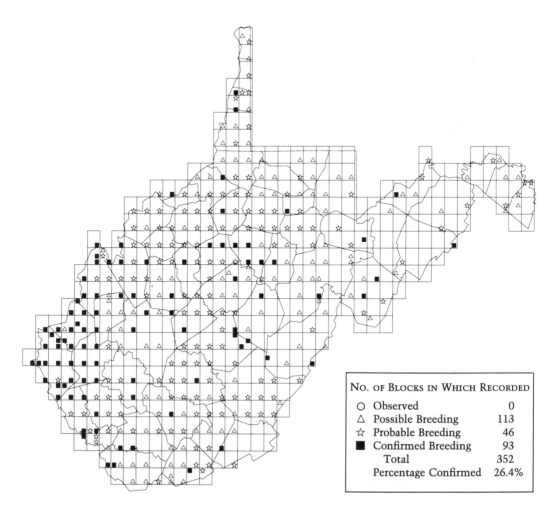

NO. OF BLOCKS IN WHICH RECORDED	
O Observed	0
△ Possible Breeding	113
☆ Probable Breeding	46
■ Confirmed Breeding	93
Total	352
Percentage Confirmed	26.4%

Solitary Vireo *Vireo solitarius*

The Solitary Vireo is generally a bird of the northern coniferous or mixed coniferous-hardwoods forests. Its northeastern range extends from central Manitoba, central Ontario, southern Quebec, and New Brunswick, south to northeastern Minnesota, central Wisconsin, northern Illinois, northern Ohio, northcentral Pennsylvania, and northern New Jersey, with a southward extension along the Appalachian highlands to northern Georgia (AOU 1983). The BBS data for this region show that from 1966 through 1978, populations increased at the rate of 6.1 percent per year ($p < .01$), but since 1978 no significant change has occurred (Robbins, Sauer, Greenberg, and Droege 1989).

In West Virginia, Atlas workers found that the Solitary Vireo has its main distribution at elevations above 300 meters in the Allegheny Mountains. Here it occupies the pure spruce forest or the mixed spruce-hardwoods; a few are also found in the northern hardwoods stands. In these last two forest types, the species competes with the Red-eyed Vireo. The Solitary Vireo also occurs in hardwoods-hemlock or even pure hardwoods on the higher ridges of the Ridge and Valley Province and in the southern part of the Western Hills Section. In these locations it does not reach the population densities found in the Alleghenies. West Virginia populations have undergone no significant change during the BBS period.

The Solitary Vireo suspends its nest from a fork on a horizontal branch 1 to 6 meters above the ground. The nest is built in the typical vireo cup shape.

Some reports of this species may be in error since occasionally the song is difficult to distinguish from that of the more abundant Red-eyed Vireo.

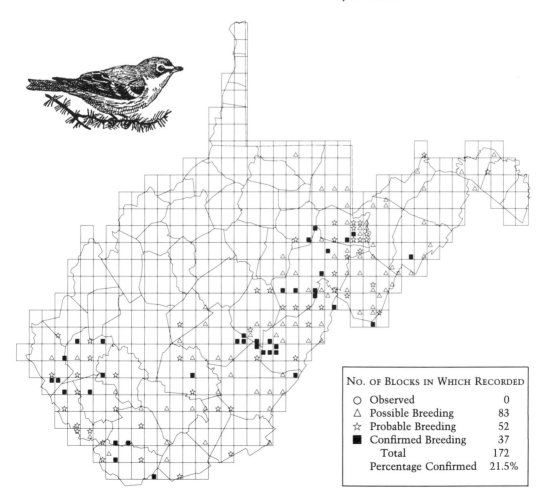

NO. OF BLOCKS IN WHICH RECORDED	
○ Observed	0
△ Possible Breeding	83
☆ Probable Breeding	52
■ Confirmed Breeding	37
Total	172
Percentage Confirmed	21.5%

Yellow-throated Vireo *Vireo flavifrons*

The YELLOW-THROATED VIREO is a widespread species that never attains very high populations. It occupies mature deciduous forest at the lower elevations from southern Minnesota, central Wisconsin, southern Ontario, and northern New Hampshire, south to central Florida (AOU 1983). It thus overlaps both geographically and ecologically with its more common congener, the Red-eyed Vireo. During the BBS period, 1966 through 1987, populations remained essentially constant (Robbins, Sauer, Greenberg, and Droege 1989).

The Atlas data show that the center of distribution of the Yellow-throated Vireo in West Virginia is in the Ohio Valley and the western part of the Western Hills Region. It is also fairly widespread in the valleys of the Potomac drainage. It is not widespread in the Allegheny Mountains Region and, oddly, was not found in the extreme east of the Eastern Panhandle.

BBS data show that, like the national populations of Yellow-throated Vireos, West Virginia populations were essentially unchanged over the period from 1966 through 1989. These data also give an average of 2.1 birds per route, as compared with 23.9 for the Red-eyed Vireo (Robbins, Bystrak, and Geissler 1986). Despite the results, there is some indication that this species, which winters in the Neotropics, may be declining in numbers (Hall pers. obs.).

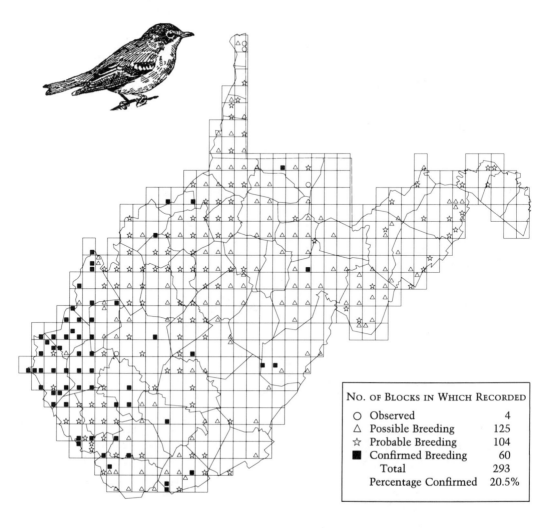

No. of Blocks in Which Recorded	
O Observed	4
△ Possible Breeding	125
☆ Probable Breeding	104
■ Confirmed Breeding	60
Total	293
Percentage Confirmed	20.5%

Warbling Vireo *Vireo gilvus*

To most birders, the WARBLING VIREO, a colorless, nondescript species, is just a pleasant voice coming from the tops of streamside trees. It occurs in suitable habitat from northern Minnesota, southern Ontario, southern Maine, and New Brunswick, south to central Mississippi, southeastern Tennessee, western North Carolina, and Virginia (AOU 1983). BBS data show essentially stable populations since 1966 (Robbins, Sauer, Greenberg, and Droege 1989).

In West Virginia, the Atlas workers found the main range to be in the Ohio and Potomac valleys. The Warbling Vireo was rarely encountered in the highlands of the Allegheny Mountains or the higher parts of the Western Hills. It favors the edges of mature forests with tall trees, and most records were in riparian situations, particularly where the streams were lined with willows

and sycamores. The species also occurs in public parks and occasionally in large trees around a farmhouse. The nest is a typical vireo cup, placed in a fork near the end of a branch and usually at an elevation of 6 to 30 meters above ground. The low percentage of "confirmed" records attests to the difficulty of finding these well-concealed nests.

West Virginia Warbling Vireo populations have declined at an alarming rate of 3.7 percent per year (p <.01) from 1966 through 1989, but the decline seems to have leveled off in the 1980s (BBS data). Since forest fragmentation would favor this species, the decline may be due to deforestation in the Central American wintering grounds, although increased cowbird parasitism is surely important for this edge-dwelling species. This vireo should be watched carefully, although it is not yet at a danger point.

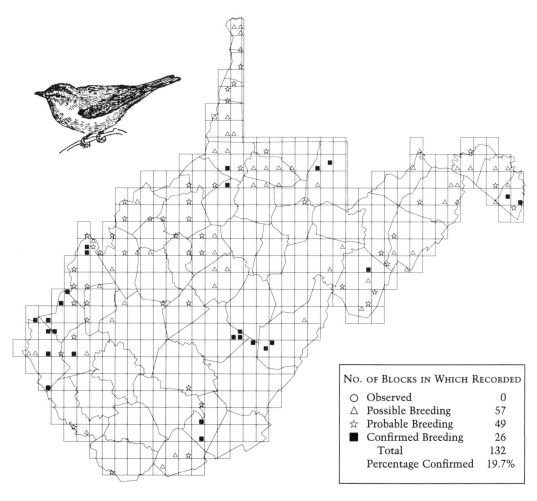

NO. OF BLOCKS IN WHICH RECORDED	
○ Observed	0
△ Possible Breeding	57
☆ Probable Breeding	49
■ Confirmed Breeding	26
Total	132
Percentage Confirmed	19.7%

Red-eyed Vireo *Vireo olivaceus*

The incessant song of the RED-EYED VIREO is the most characteristic sound in West Virginia's woodlands in the summer. This vireo sings through most of the day and on through the hot days of July. It nests throughout eastern North America from the limit of trees in the north, south to central Florida. There is probably no wooded tract of sufficient size in this whole area that lacks at least one pair of Red-eyed Vireos; Robbins, Dawson, and Dowell (1989) found it in woodlots as small as 0.8 to 1.2 hectares. From 1966 to 1978 the BBS data for the Northeast showed an increase of 2.8 percent per year (p <.01), but since that time populations have been stable (Robbins, Sauer, Greenberg, and Droege 1989).

The Atlas data confirmed that the Red-eyed Vireo nests throughout West Virginia, where it was the sixth most frequently reported species. Those few blocks that contain no records for this species were probably not covered very well. The percentage of "confirmed" records is perhaps not as high as it might have been, since nests are hard to find in the leafy canopy. The nests are compact cups placed in a fork near the end of a branch, usually 7 to 20 meters above ground. If Atlas volunteers had toured the blocks in the fall after the leaves were down, many used nests could have been found.

This species occurs in all types of mature forest except the pure spruce forest (Hall 1983), and it is probably the most numerous woodland bird in the state. An average of 23.7 birds per route was observed for this vireo on the West Virginia BBS routes (Robbins, Bystrak, and Geissler 1986). Populations in the state have shown no significant trends in the years from 1966 through 1989 (BBS data).

Forest fragmentation is not as serious for this species as it is for some. However, fragmentation does lead to increased cowbird parasitism, which will be important in the future of this species.

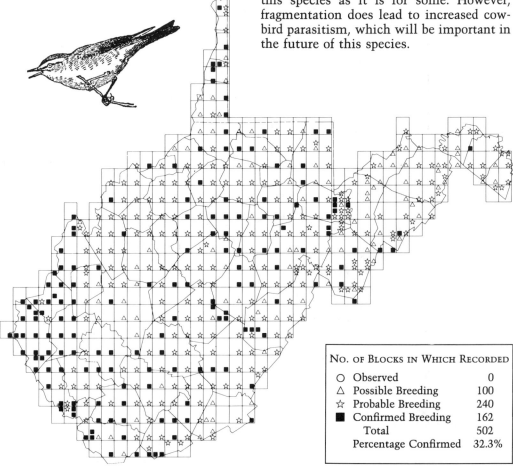

No. of Blocks in Which Recorded	
O Observed	0
△ Possible Breeding	100
☆ Probable Breeding	240
■ Confirmed Breeding	162
Total	502
Percentage Confirmed	32.3%

Blue-winged Warbler *Vermivora pinus*

The BLUE-WINGED WARBLER is an uncommon species of the Midwest. Its range extends from southern Minnesota, southern Wisconsin, southern Michigan, northern Ohio, southern Ontario, central New York, southern Vermont, and Massachusetts, south to southern Missouri, northern Alabama, northern Georgia, western North Carolina, and northern Virginia (AOU 1983). In this region it is an inhabitant of second-growth woodlands and brushy areas such as power line right-of-ways. The Blue-winged Warbler has entered the eastern portion of this range in relatively recent times. BBS data show the species to have remained essentially constant in population, although the range expansion continues (Robbins, Sauer, Greenberg, and Droege 1989).

Hall (1983) indicated that the principal West Virginia range was in the counties along the Ohio River, where the bird nested on the ridges just back from the river. It also occurs in the lower Kanawha and Monongahela drainages. He also cited summer records from 13 other counties. The Atlas results confirmed that the species occurred throughout the Ohio River drainage, covering most of the Western Hills country. There were also a few records at low elevations in the Allegheny Mountains and in the Eastern Panhandle.

This warbler's habitat is second growth of all types of hardwoods forest. The species tolerates older growth better than does the Golden-winged Warbler, which is one of the factors leading to the gradual displacement of the Golden-winged by the Blue-winged Warbler. West Virginia populations have not shown a significant trend in the period from 1966 through 1989 (BBS data).

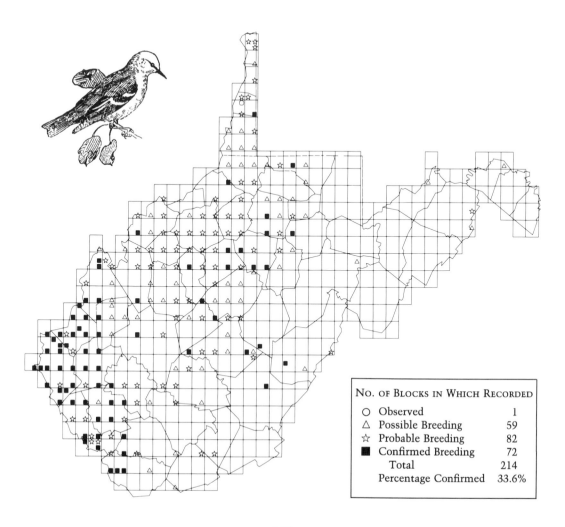

No. of Blocks in Which Recorded	
O Observed	1
△ Possible Breeding	59
☆ Probable Breeding	82
■ Confirmed Breeding	72
Total	214
Percentage Confirmed	33.6%

Vermivora Hybrids
("Brewster's" and "Lawrence's" Warblers)

In the regions in which they are sympatric, the Golden-winged and Blue-winged warblers hybridize readily. The first-generation cross between two parents of pure lineage is known as the "Brewster's" Warbler. It resembles a Blue-winged Warbler, with mostly white underparts.

Another form, thought to be one of the possible results produced by matings of a Brewster's Warbler with one of the parent species or with another Brewster's, is called "Lawrence's" Warbler. This bird resembles a Golden-winged Warbler, with yellow underparts. It is the much rarer of the two forms.

Other birds of obvious mixed ancestry occur, sharing the characteristics of the parent species in varying degrees. In some areas, a large percentage of birds may show some intergradation of characteristics, although this may not always be apparent unless the bird can be held in the hand and examined.

West Virginia Atlas workers found Brewster's Warblers in seven blocks. In two blocks, "confirmed" records were of sightings of adults carrying food. There was one sighting of a Lawrence's Warbler.

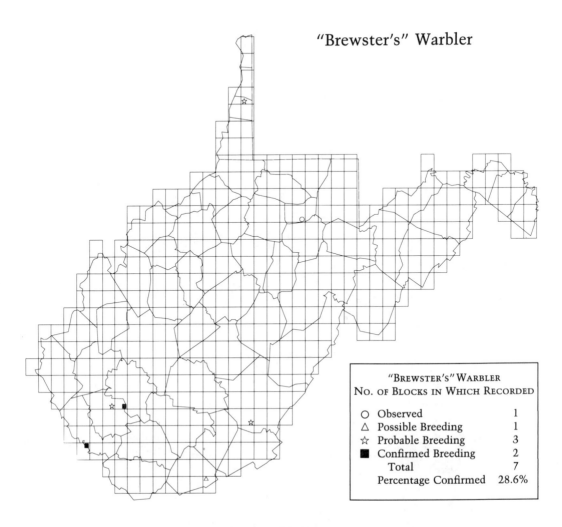

"Brewster's" Warbler

"Brewster's" Warbler
No. of Blocks in Which Recorded

O	Observed	1
△	Possible Breeding	1
☆	Probable Breeding	3
■	Confirmed Breeding	2
	Total	7
	Percentage Confirmed	28.6%

Vermivora Hybrids
("Brewster's" and "Lawrence's" Warblers)

"Lawrence's" Warbler

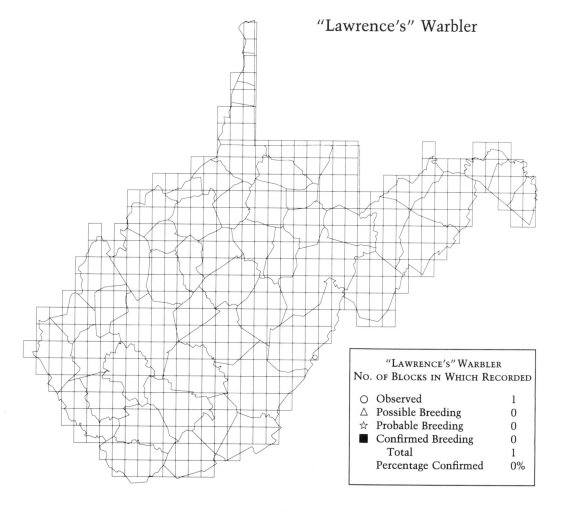

"LAWRENCE'S" WARBLER	
No. of Blocks in Which Recorded	
○ Observed	1
△ Possible Breeding	0
☆ Probable Breeding	0
■ Confirmed Breeding	0
Total	1
Percentage Confirmed	0%

Golden-winged Warbler *Vermivora chrysoptera*

The GOLDEN-WINGED WARBLER is an uncommon species nesting from northern Wisconsin, northern Michigan, southern Ontario, northern New York, and southern New England, south to northern Illinois, northern Indiana, southern Ohio, and southeastern Pennsylvania, with an extension south to Tennessee and Georgia in the Appalachians (AOU 1983). It nests in low second-growth and open woodlands, and it especially favors power line right-of-ways. In the period from 1966 through 1978, eastern populations declined at an annual rate of 2.2 percent (*p* <.05), and from 1978 through 1987 the decline was 1.9 percent per year (not significant at the *p* = 0.05 level) (Robbins, Sauer, Greenberg, and Droege 1989).

Hall (1983) listed summer records for 27 West Virginia counties, whereas the Atlas fieldwork located the Golden-winged Warbler in approximately 40 counties. It was largely missing from most of the Ohio Valley and of very local occurrence in the Ridge and Valley Region. The species was most common at middle elevations on the western side of the mountains and in the Western Hills Region. The birds disappear from their brushy second-growth habitat at a fairly early stage in the successional sequence.

West Virginia populations have declined at an annual rate of 4.8 percent (*p* <.05) in the period from 1966 to 1987 (BBS data). This decline can be traced partly to the loss of suitable habitat, but another important factor has been the increase in Blue-winged Warbler numbers and range. This species seems to displace the Golden-wing and is also able to tolerate older stages of plant succession, remaining in places the Golden-wing has left. There may also be genetic swamping as the two species hybridize.

The Golden-winged Warbler seems now to be limited to the higher elevations of West Virginia. Although power line right-of-ways are prime habitat for this species, increased use of defoliant sprays reduces the amount of this habitat available.

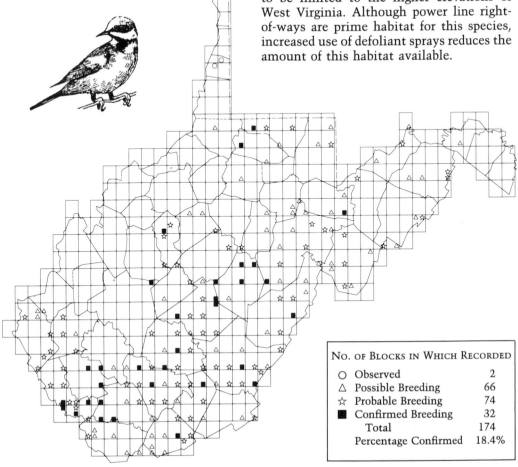

NO. OF BLOCKS IN WHICH RECORDED	
○ Observed	2
△ Possible Breeding	66
☆ Probable Breeding	74
■ Confirmed Breeding	32
Total	174
Percentage Confirmed	18.4%

Nashville Warbler *Vermivora ruficapilla*

The NASHVILLE WARBLER, a bird of the northern forest, reaches its southernmost station in West Virginia. It inhabits a variety of both coniferous and deciduous second-growth woodland as well as the forest edge of bogs. It ranges from the limit of trees to the North, south to northern Minnesota, southern Wisconsin, northeastern Illinois, southern Michigan, southern Ontario, northern Pennsylvania, northeastern New Jersey, and southern New England, with an extension along the Allegheny Mountains in Pennsylvania that barely reaches northern West Virginia and western Maryland (AOU 1983). Eastern populations grew at a small rate (not significant at the $p = 0.05$ level) in the period between 1978 and 1987 (Robbins, Sauer, Greenberg, and Droege 1989).

In West Virginia, Atlas records for the Nashville Warbler are concentrated near the bogs of the highlands of Tucker and Grant counties, essentially on the Allegheny Front. There was a historic population on Canaan Mountain, where the first and only nest was discovered in 1951 (Hall 1983), but with the growth of the spruce trees at that station, this population seems to have disappeared. There have been occasional summer records from other places in the Allegheny Mountains, such as Cranberry Glades, Pocahontas County, and Cranesville Swamp, Preston County. Atlas workers reported one "confirmed" record in Tucker County in 1985. A Nashville Warbler nesting was established in the 1950s on the Maryland side of the Cranesville Swamp, and the Maryland Atlas project recorded it at several places in Garrett County (Maryland Atlas data).

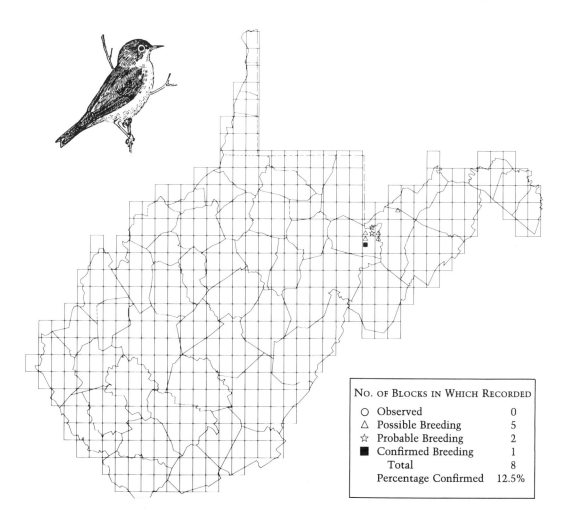

NO. OF BLOCKS IN WHICH RECORDED	
○ Observed	0
△ Possible Breeding	5
☆ Probable Breeding	2
■ Confirmed Breeding	1
Total	8
Percentage Confirmed	12.5%

Northern Parula *Parula americana*

The diminutive NORTHERN PARULA is a widespread but never common species in the eastern United States. It nests from central Ontario, southern Quebec, and the Maritimes, south to the Gulf Coast and northern Florida (AOU 1983), although there are many gaps in that range. Eastern populations declined at an annual rate of 2.1 percent ($p < 0.01$) from 1978 through 1987 (Robbins, Sauer, Greenberg, and Droege 1989).

Hall (1983) suggested that the Northern Parula might nest in every county in West Virginia, and the Atlas workers confirmed this. With a few gaps, some of which are probably artifacts of coverage, this species occurs throughout the state. The gap in the Northern Panhandle may be real, since the Pennsylvania Atlas did not find it in the neighboring area to the east, although the Ohio Atlas project found it in a few blocks to the west and north (Peterjohn and Rice 1991). Populations are never large, and in any one area the distribution is very spotty.

The Northern Parula occurs in the mature stages of all forest types, including limited numbers in the edge of the spruce forest, with the largest populations being in hemlock-hardwoods stands and in sycamore bottomlands. In recent years it has appeared widely in suburban residential areas, where planted conifers such as blue spruce and hemlock have reached large size. West Virginia lacks both *Usnea* lichen and Spanish moss, the nesting substrates the species uses in other parts of its range, so nests are made in situations where fallen leaves and flood or other debris have been trapped in trees. Such nests are not easy to find.

The BBS data for West Virginia show no significant trends over the period 1967 through 1989, but there is a suggestion of increase in the last 10 years. Robbins, Dawson, and Dowell (1989) postulated a minimum area for the species at 500 hectares, with 10 hectares as the smallest area of occurrence. These data, however, take no account of the recent surge in numbers of ornamental conifers.

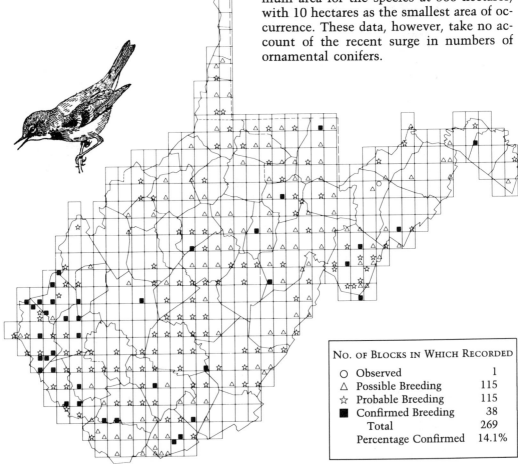

NO. OF BLOCKS IN WHICH RECORDED	
○ Observed	1
△ Possible Breeding	115
☆ Probable Breeding	115
■ Confirmed Breeding	38
Total	269
Percentage Confirmed	14.1%

Yellow Warbler *Dendroica petechia*

The YELLOW WARBLER, with the widest distribution of any member of the warbler family, is found throughout eastern North America from the limit of trees in central Canada, south to northern Arkansas, northern Mississippi, central Alabama, and central Georgia (AOU 1983). It occurs in second-growth woodland, hedgerows, marshland edges, farmlands, and gardens. Eastern populations showed fluctuations but no general trends during the period from 1966 through 1978 (Robbins, Bystrak, and Geissler 1986).

In West Virginia, the Yellow Warbler is the most common warbler and is found throughout the state wherever suitable habitat exists, even at high elevations. It occurs in house yards, cemeteries, abandoned fields, and other edge situations. The species does not enter the spruce forest, but can be found at high elevations in streamside and bog situations. The BBS data from 1966 to 1989 show a very slight, nonsignificant decrease.

This warbler places its nest, which is a compact cup of interwoven plant fibers, in an upright fork or crotch of a shrub or tree 1 to 4 meters above the ground. The species is one of the few that have adapted to minimize nest parasitism by the Brown-headed Cowbird; it often buries the cowbird eggs beneath a layer of nesting material where they fail to hatch.

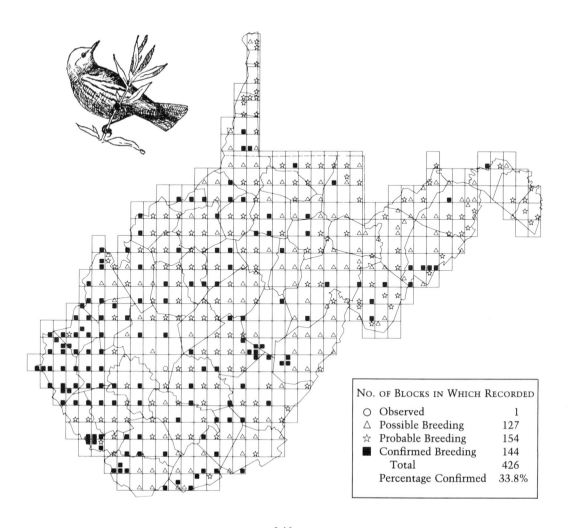

No. of Blocks in Which Recorded	
○ Observed	1
△ Possible Breeding	127
☆ Probable Breeding	154
■ Confirmed Breeding	144
Total	426
Percentage Confirmed	33.8%

Chestnut-sided Warbler *Dendroica pensylvanica*

The CHESTNUT-SIDED WARBLER is a bird of the early, brushy stages of plant succession, and populations vary widely depending on the amount of habitat available. The species ranges from the limit of trees in the North, south to southern Wisconsin, southern Michigan, northern Ohio, southern Pennsylvania, and northern New Jersey, with an extension along the Appalachian highlands south to northern Georgia and western South Carolina (AOU 1983). BBS data for the years 1966 to 1978 indicated an annual increase of 2.2 percent ($p <0.01$), but from 1978 to 1987, populations declined at an annual rate of 3.8 percent ($p <0.01$) (Robbins, Sauer, Greenberg, and Droege 1989).

In West Virginia the Chestnut-sided Warbler is found throughout the Allegheny Mountains Region at elevations usually above 500 meters. It also occurs in the higher portions of the southern Western Hills Region and on some of the higher ridges in the Ridge and Valley Region. This warbler inhabits the brushy stages of succession, where it places its nest, a loosely woven cup, close to the ground.

In presettlement times, the Chestnut-sided Warbler was very rare in West Virginia. As the forest was removed in the late nineteenth and early twentieth centuries, the species increased until it was one of the most common birds in the highlands. Lately, as succession has advanced, it has become less common, although recent lumbering activity in the Monongahela National Forest has opened up more suitable habitat. West Virginia BBS data show a 12.4 percent per year ($p <0.05$) increase in the period from 1980 to 1989.

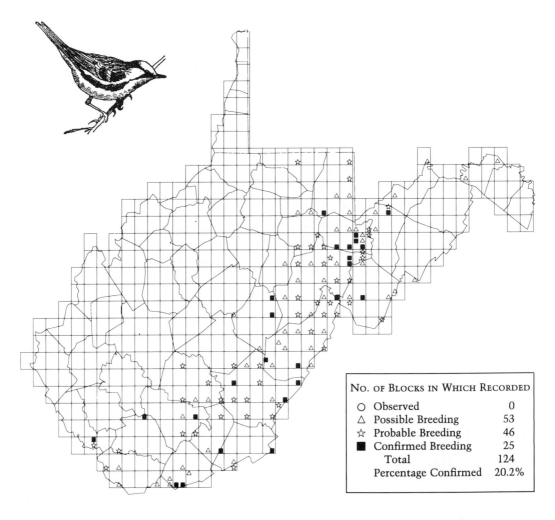

NO. OF BLOCKS IN WHICH RECORDED	
○ Observed	0
△ Possible Breeding	53
☆ Probable Breeding	46
■ Confirmed Breeding	25
Total	124
Percentage Confirmed	20.2%

Magnolia Warbler *Dendroica magnolia*

The attractive MAGNOLIA WARBLER is another member of the so-called northern element in the West Virginia avifauna. Its normal range is from the northern edge of the boreal forest, south to northern Minnesota, northern Wisconsin, central Michigan, southern Ontario, northern Pennsylvania, northeastern New Jersey, and Connecticut, with a southward extension along the Alleghenies to West Virginia, western Maryland, and western Virginia (AOU 1983). In recent years there have been summer records as far south as North Carolina, and an isolated population occurs in Hocking County in southern Ohio (Peterjohn and Rice 1991). Eastern populations remained essentially constant during the BBS years (Robbins, Bystrak, and Geissler 1986).

Atlas workers found the Magnolia Warbler limited to the higher Allegheny Moun-

tains in West Virginia. Its distribution is essentially the same as that mapped by Hall (1983), who stated that it was normally found above 900 meters. The Magnolia Warbler occurred in large numbers in young spruce forest and was present in mature forest if there was an understory. It was also found in the early successional stages of the mixed spruce-hardwoods forest, and in lesser numbers in the northern hardwoods forest. The BBS data showed that West Virginia populations had increased 7.0 percent per year ($p < 0.01$) in the period from 1966 to 1989, but only five routes reported the species.

The low percentage of "confirmed" Atlas records for the Magnolia Warbler probably reflects the difficulty of finding its nest, which is commonly built in dense Christmas-tree size conifers (Harrison, 1975).

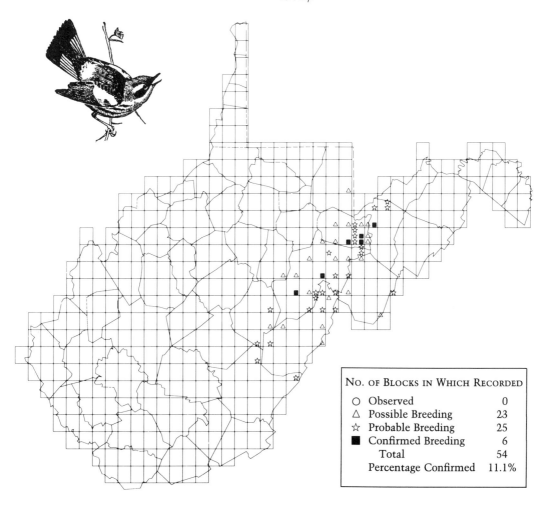

NO. OF BLOCKS IN WHICH RECORDED	
O Observed	0
△ Possible Breeding	23
☆ Probable Breeding	25
■ Confirmed Breeding	6
Total	54
Percentage Confirmed	11.1%

Black-throated Blue Warbler *Dendroica caerulescens*

The quietly beautiful BLACK-THROATED BLUE WARBLER is a common resident of the northern forest wherever there is thick undergrowth. It typically inhabits rhododendron tangles in the part of its range where this plant occurs. The nesting range extends from western Ontario, southern Quebec, and the Maritimes, south to northern Minnesota, northern Wisconsin, northern Michigan, southern Ontario, northern Pennsylvania, and southern New York, with an extension along the Appalachian highlands south to northern Georgia and western North Carolina (AOU 1983). Populations have remained fairly constant during the BBS period (Robbins, Bystrak, and Geissler 1986).

Hall (1983) indicated that the Black-throated Blue Warbler was limited to the Allegheny Mountains Region at elevations above 600 meters. The Atlas results confirmed this distribution, but they also showed a population in the higher parts of the southern Western Hills Region, where the northern hardwood forest is present. As mentioned above, this warbler is often found in rhododendron-choked hollows, but it may occur in young spruce stands and occasionally at lower elevations in cool hemlock-covered ravines. As with several of the other warblers inhabiting a dense undergrowth, its nests are difficult to find, accounting for the low percentage of "confirmed" records.

The BBS data indicate that West Virginia populations increased at an annual rate of 14.8 percent ($p < 0.01$) between 1980 and 1989, but only six BBS routes recorded it during that period. This species appears to be very area-sensitive, since Robbins, Dawson, and Dowell (1989) did not find it on tracts of less than 1,000 hectares. Continued forest fragmentation may represent a threat to the Black-throated Blue Warbler populations.

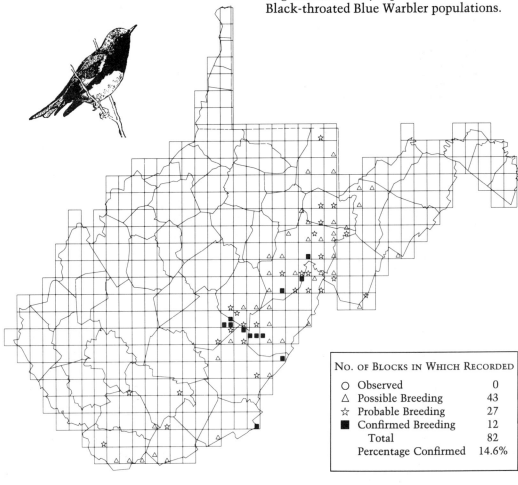

No. of Blocks in Which Recorded	
○ Observed	0
△ Possible Breeding	43
☆ Probable Breeding	27
■ Confirmed Breeding	12
Total	82
Percentage Confirmed	14.6%

Yellow-rumped Warbler *Dendroica coronata*

The YELLOW-RUMPED WARBLER, which was long familiar only as an abundant migrant and occasional winter resident, has been added to the West Virginia breeding avifauna only in the last 15 years.

As previously understood, this warbler's range was from the limit of trees in the North, south to northern Minnesota, northern Wisconsin, northern Michigan, southern Ontario, northern New York, northeastern Pennsylvania, and southern New England (AOU 1983). The bird nests in coniferous woodland or mixed coniferous-hardwoods forest, where it places its nest on a horizontal branch at heights ranging from 1 to 15 meters. In the period from 1966 to 1978, eastern populations showed a steady significant increase (Robbins, Bystrak, and Geissler 1986), and its range has been shifted southward.

The Pennsylvania Atlas project found the Yellow-rumped Warbler widespread throughout the northern half of that state (Brauning 1992).

The first summer records for West Virginia came from Gaudineer Knob on the Randolph County-Pocahontas County line in 1975. Since that time, the bird has appeared in almost every location where red spruce of suitable age is growing. Atlas workers found it in Tucker, Pendleton, Randolph, and Pocahontas counties. The first "confirmed" nesting was made on Spruce Knob, Pendleton County, in 1987 (Eddy 1988). In 1982 the population on Spruce Knob was measured at 41 males per 100 hectares (Smith and Hall 1983). This species will probably continue to extend its range in the near future.

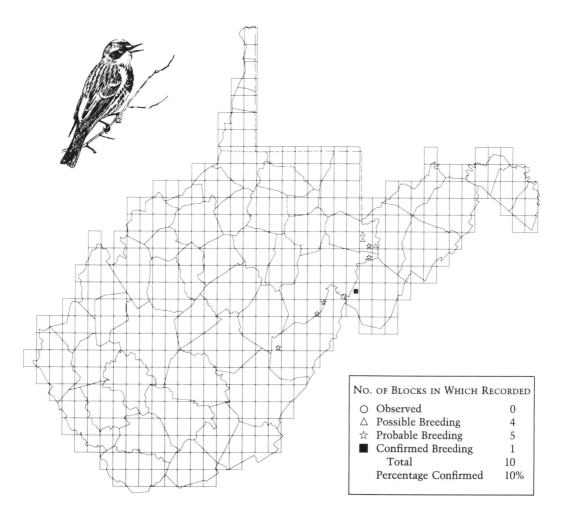

NO. OF BLOCKS IN WHICH RECORDED	
○ Observed	0
△ Possible Breeding	4
☆ Probable Breeding	5
■ Confirmed Breeding	1
Total	10
Percentage Confirmed	10%

Black-throated Green Warbler *Dendroica virens*

The dreamy song of the BLACK-THROATED GREEN WARBLER is typical of summer in the northern forests. The species nests from central Ontario, southern Quebec, southern Labrador, and Newfoundland, south to northern Minnesota, northern Wisconsin, southern Michigan, northeastern Ohio, central Pennsylvania, and southern New York, with a projection south through West Virginia to northern Alabama and northern Georgia (AOU 1983). Over most of this range, it inhabits mature coniferous forest or mixed hardwoods-coniferous forest.

Populations of the Black-throated Green Warbler are subject to oscillations of large magnitude. Between 1978 and 1987, the eastern population declined at an annual rate of 3.3 percent ($p < 0.01$) (Robbins, Sauer, Greenberg, and Droege 1989).

Hall (1983) suggested that the Black-throated Green Warbler nested in every county in West Virginia except those immediately along the Ohio River and the two easternmost counties. The Atlas results partially confirmed this, but the bird was not found in a wide belt of Western Hills counties. It was most widely distributed in the Allegheny Mountains and the hills along the state's southwestern border. An isolated population is located in Hancock County and some adjoining counties in Ohio (Peterjohn and Rice 1991). An isolated population also occurs in southeastern Ohio. Although this warbler nests in the mixed hardwoods-spruce forest, it was also commonly found in some places in northern hardwoods stands that had no spruce. In the southwestern part of the state it occurred on ridges covered with mixed mesophytic forest. Along the western edge of the mountains, it was often found in cool ravines containing some hemlock. The isolated populations were usually in this sort of situation.

Although BBS data indicate that West Virginia populations have increased in the period from 1966 to 1989 at the rate of 3.7 percent per year ($p < 0.05$), examination of annual data shows that the population of this species varies between wide limits in a nonperiodic fashion (Hall 1984).

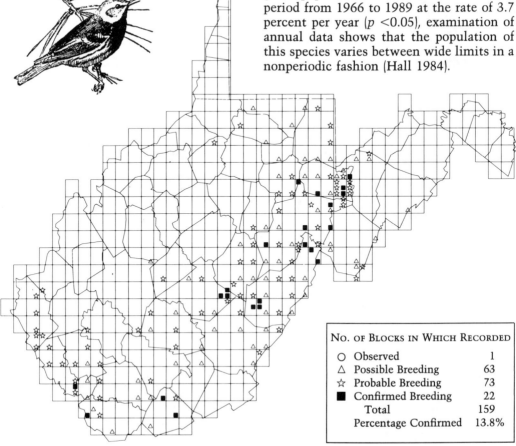

No. of Blocks in Which Recorded	
O Observed	1
△ Possible Breeding	63
☆ Probable Breeding	73
■ Confirmed Breeding	22
Total	159
Percentage Confirmed	13.8%

Blackburnian Warbler *Dendroica fusca*

During its migration, the vividly colored BLACKBURNIAN WARBLER is a familiar sight to bird students, but it is much less familiar in its breeding season. Usually classified as a Canadian Zone species, this warbler nests from central Ontario, central Quebec, Labrador, and Newfoundland, south to central Wisconsin, northern Michigan, northeastern Ohio, and northern Pennsylvania, with a southward projection along the higher Appalachians to northern Georgia (AOU 1983). It nests in the coniferous forest, particularly the spruce-fir forest in the southern part of its range. Occasionally it will appear at lower elevations.

The West Virginia population of Blackburnian Warblers is generally restricted to the spruce forest and mixed spruce-hardwoods forest in the higher Allegheny Mountains, usually above 900 meters. The species does occur in the oak-hickory-pine forest (see, for example, the Morgan County records on the Atlas map). In these situations, this species may occur as low as 600 meters.

This warbler usually places its nest high in dense conifer foliage, usually in a tall spruce tree. Nests are thus very difficult to find, and most of the "confirmed" Atlas records are of sightings of fledged young.

West Virginia populations increased at an annual rate of 2.1 percent ($p < 0.05$) in the period from 1966 to 1989 (BBS data), but the number of routes was low. From 1980 to 1989, the BBS data showed a decrease of 4 percent per year (not significant at the $p = 0.05$ level). When coupled with other census data, this figure suggests that a decline may be starting. Breeding densities in the virgin spruce-hardwoods area in Randolph County have remained essentially constant from 1948 to 1988. Because the spruce forest in West Virginia is under environmental stress, the Blackburnian Warbler is one of the species that should be monitored carefully.

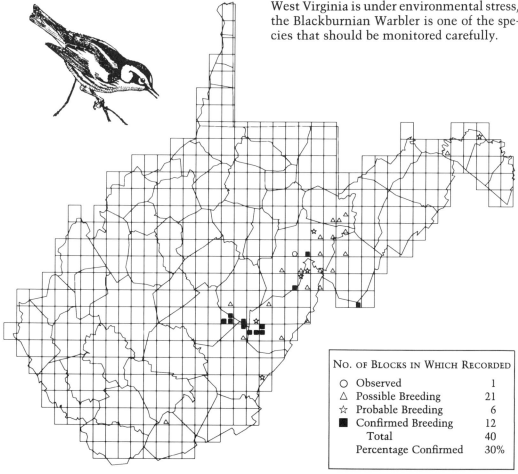

NO. OF BLOCKS IN WHICH RECORDED	
○ Observed	1
△ Possible Breeding	21
☆ Probable Breeding	6
■ Confirmed Breeding	12
Total	40
Percentage Confirmed	30%

Yellow-throated Warbler *Dendroica dominica*

The YELLOW-THROATED WARBLER is a relatively new addition to the birdlife of West Virginia and its neighboring states. Todd (1940) had no records for Western Pennsylvania; Hicks (1935) had only a few breeding records for Ohio; and Brooks (1944) had only scattered records for West Virginia. There is evidence (Peterjohn and Rice 1991) that in the nineteenth century the Yellow-throated Warbler was much more widespread in this general region. In the 1940s this species was primarily a bird of the southern states, nesting in mature bottomland forest, particularly sycamore woodland, and also in the open southern pine forests. Since at least the 1950s, it has been expanding its range northward, and it now occurs widely in southwestern Pennsylvania and in southern and eastern Ohio. In the years 1966 to 1978, eastern populations increased at a rate of 2.0 percent per year (*p* <0.05).

West Virginia Atlas workers found the Yellow-throated Warbler to be widespread along the Ohio River and its tributaries. It occurs sparingly in the eastern part of the Western Hills and is absent from the Allegheny Mountains, although the 1989 Brooks Bird Club Foray located one at high elevation along Shavers Fork. Hall (1983) mentioned a population along the Shenandoah and Potomac rivers in eastern West Virginia, but no Atlas records were reported there, possibly due to limitations of the Atlas sample. (See "Limitations and Biases" in the Introduction.) This species is most common in mature bottomland hardwoods, although in the Charleston area it occurs on the ridges covered with scrub pine. BBS data show that West Virginia populations increased at the amazing rate of 13.6 percent per year (*p* <0.01) during the years of 1966 through 1987, but a downward trend has been noted in the 1980s.

Yellow-throated Warblers place their nests in the canopies of tall trees, where they are not easily found except during the nest-building period before the trees leaf out. The percentage of "confirmed" Atlas records of breeding for this species is thus exceptionally good.

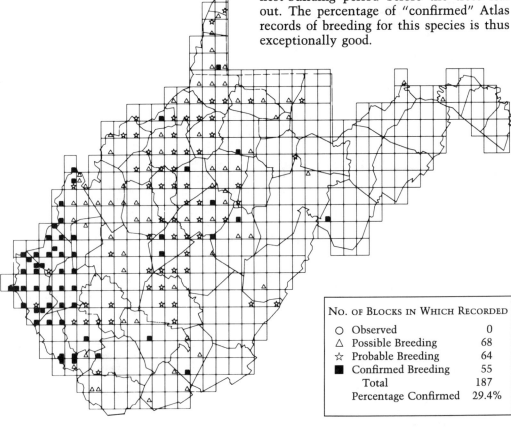

NO. OF BLOCKS IN WHICH RECORDED	
○ Observed	0
△ Possible Breeding	68
☆ Probable Breeding	64
■ Confirmed Breeding	55
Total	187
Percentage Confirmed	29.4%

Pine Warbler *Dendroica pinus*

If ever a species was aptly named, it is the PINE WARBLER. Even in migration this bird is seldom if ever seen anywhere except in a pine woods, where its song can often be mistaken for that of either a Chipping Sparrow or a Worm-eating Warbler or, at high elevations, even a Dark-eyed Junco.

The continental range for this warbler extends from northern Wisconsin, northern Michigan, central Ontario, and central Maine, south to the Gulf Coast and Florida (AOU 1983). In this extensive range, it occurs locally wherever there are pines, such as the southern pine forest or the jack pine forest of the North. Because the species winters in the southern United States, the population often declines after a severe winter in the South.

In West Virginia the Pine Warbler occurs in the Virginia pine woodlands of the Ridge and Valley Region and in the pine woods of southwestern West Virginia. It can also be found in the remnant white pine areas of the Upper Greenbrier Valley, but it is largely missing from the northern part of the Allegheny Mountains and the northern part of the Western Hills. BBS data show little change in population, but the number of routes reporting the species is too small for a good trend.

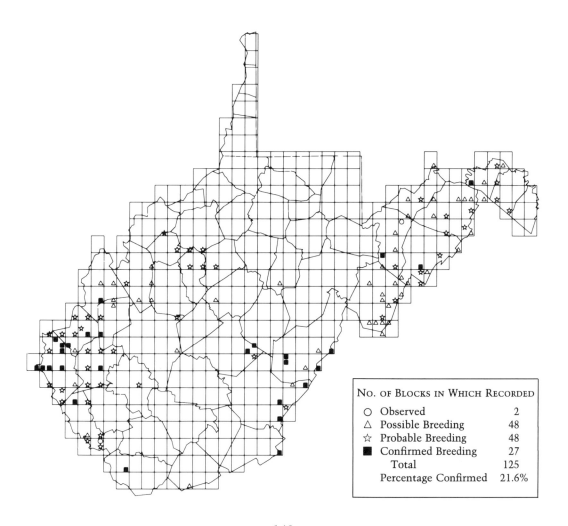

NO. OF BLOCKS IN WHICH RECORDED	
○ Observed	2
△ Possible Breeding	48
☆ Probable Breeding	48
■ Confirmed Breeding	27
Total	125
Percentage Confirmed	21.6%

Prairie Warbler *Dendroica discolor*

The PRAIRIE WARBLER is a bird of the young pine forests and other brushy scrub areas from central Wisconsin, southern Michigan, southern Pennsylvania, southeastern New York, and southern New England, south to southern Florida (AOU 1983). Eastern populations declined at the rate of 3.7 percent per year (p <0.01) in the period 1966 to 1978, but since 1978 there has been no significant decline (Robbins, Sauer, Greenberg, and Droege 1989).

The Atlas workers found the Prairie Warbler widespread in the Western Hills and Ridge and Valley regions. It occurred only locally in the higher Allegheny Mountains. Although the species has been observed at elevations as high as 1,100 meters, it seems to have disappeared from these higher sites in recent years, possibly as a result of maturation of what was once suitable habitat. As pointed out by Hall (1983), the Prairie War-

bler's range in West Virginia has increased greatly in the last 50 years. In Ohio, the species appears limited mainly to the Appalachian Plateau; the distribution pattern in Pennsylvania is similar to that in West Virginia.

The Prairie Warbler is found widely in young second-growth hardwoods, overgrown pastures, Christmas tree plantations, and similar areas. The amount of this habitat is declining, as new trees grow in some areas and land is cleared for various purposes in other areas. Thus, despite Hall's (1983) remark that populations in recent years had increased, the BBS data show a decrease of 6.4 percent per year (p <0.01) from 1966 through 1989.

The percentage of Atlas "confirmed" records of nesting seems rather low. Prairie Warbler nests, built in shrubs or in the lower boughs of pine or cedar trees, are not especially hard to find.

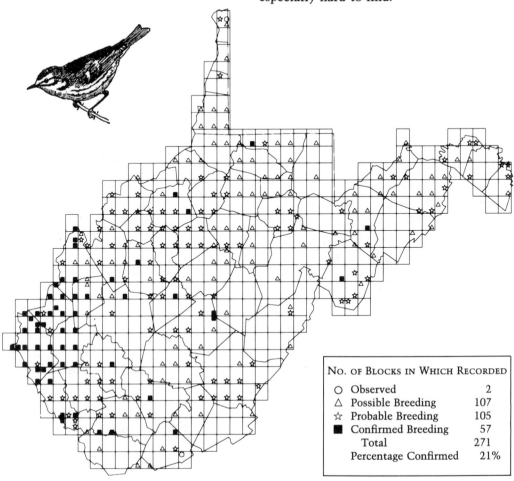

NO. OF BLOCKS IN WHICH RECORDED	
O Observed	2
△ Possible Breeding	107
☆ Probable Breeding	105
■ Confirmed Breeding	57
Total	271
Percentage Confirmed	21%

Cerulean Warbler *Dendroica cerulea*

The CERULEAN WARBLER is a more common bird in West Virginia than many birders realize. Because this warbler seldom leaves the heavily leafed canopy of the mature forest, observers who do not know the bird's song will often be unaware of its presence.

The Cerulean Warbler's principal habitat is the mixed mesophytic forest and the Appalachian oak forest at elevations below 600 meters (Hall 1983). Its main continental range is in the hardwoods forest area of the Mississippi and Ohio drainages (AOU 1983). BBS data indicate that the highest populations of this species are found in West Virginia, where an average of 1.8 birds per 50-stop route were recorded for the years 1966 through 1988 (Robbins, Fitzpatrick, and Hamel 1992). The BBS data showed a decline of 3.9 percent per year (*p* <0.01) from 1966 to 1978, but since 1978 there has

been no significant trend (Robbins, Sauer, Greenberg, and Droege 1989).

The Atlas work clearly shows that in West Virginia the Cerulean Warbler is widespread as well as common in the Western Hills. It becomes scarce or missing in the Allegheny Mountains Region and reappears sparingly in the Ridge and Valley Region. In this latter area, it is limited to the river valleys. The low number of "confirmed" breeding records attests to the difficulty of finding nests in the treetops.

The Cerulean Warbler is an area-sensitive species, and Robbins, Dawson, and Dowell (1989) suggested that a minimum area of 700 hectares was needed for a viable population. The species is subject to stress at both ends of its range, with fragmentation and elimination of habitat in the North and deforestation of its winter range on the middle Andean slopes.

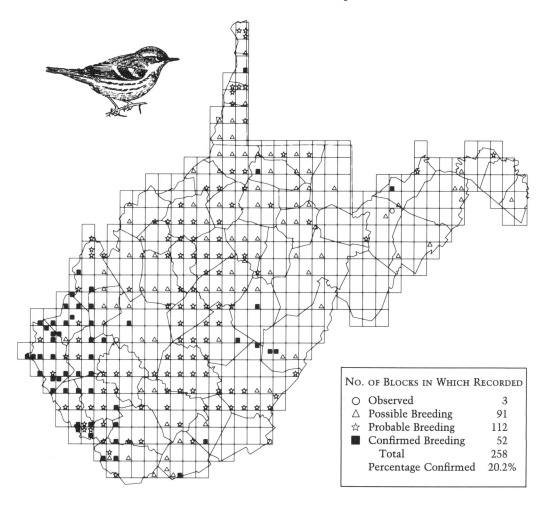

No. of Blocks in Which Recorded	
O Observed	3
△ Possible Breeding	91
☆ Probable Breeding	112
■ Confirmed Breeding	52
Total	258
Percentage Confirmed	20.2%

Black-and-white Warbler *Mniotilta varia*

The BLACK-AND-WHITE WARBLER occurs throughout the deciduous forest region, including the mixed deciduous-conifer forest, of the eastern United States. It is a denizen of the mature forest, where its inconspicuous song and its habit of creeping on trunks and branches in nuthatch fashion sometimes make it a difficult bird to locate. In the period from 1978 to 1987, eastern populations increased at an annual rate of 1.4 percent ($p < 0.05$) (Robbins, Sauer, Greenberg, and Droege 1989).

Hall (1983) remarked that the species probably nested in every county in West Virginia, but Atlas workers failed to find it in several counties. Of particular interest is its unaccountable absence from the Eastern Panhandle. Between 1966 and 1989, West Virginia populations have declined at a rate of 4.4 percent per year ($p < 0.10$) (BBS data).

The Black-and-white Warbler nests in mature deciduous forest with closed canopy and, usually, sparse ground cover (Hall 1983). The nest is placed on the ground, usually at the base of a tree or shrub. Atlas workers found very few nests, and most of the "confirmed" records were of sightings of fledged young.

The Black-and-white Warbler is an area-sensitive species, and Robbins, Dawson, and Dowell (1989) did not find it in forest tracts of less than 200 hectares.

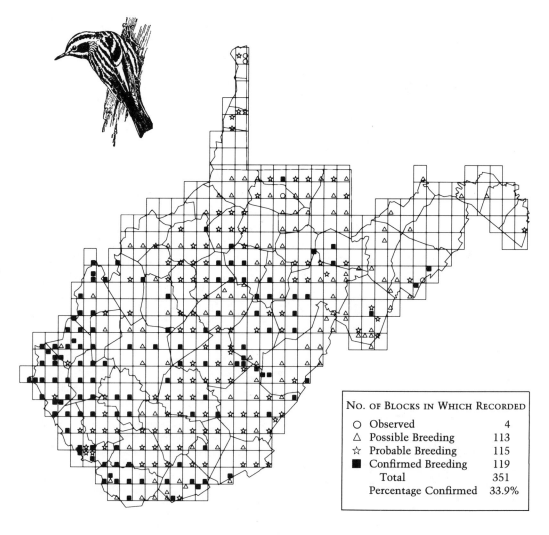

NO. OF BLOCKS IN WHICH RECORDED	
O Observed	4
△ Possible Breeding	113
☆ Probable Breeding	115
■ Confirmed Breeding	119
Total	351
Percentage Confirmed	33.9%

American Redstart *Setophaga ruticilla*

The colorful AMERICAN REDSTART is a familiar bird of the second-growth woodland that covers much of West Virginia. This species is an active feeder and can hardly be missed by an observer.

The redstart occurs throughout the forested region of the eastern United States, south to the Gulf Coast, southern Georgia, central South Carolina, and eastern North Carolina (AOU 1983). It nests in second-growth deciduous forests or in aspen groves in the coniferous forest region, and it often attains very high densities. Eastern populations have been nearly stable over the USFWS Breeding Bird Survey period.

In West Virginia, the redstart nests throughout the state wherever suitable habitat exists. It is not found in the spruce forest and is uncommon above 900 meters (Hall 1983). Its nest, which is located in a fork of a low shrub or small tree, is relatively easily found.

Peak redstart populations in the state probably occurred in the 1930s, when there was a maximum of good habitat. Since then the amount of suitable second-growth forest has declined as forests have matured. In the period spanning 1966 to 1989, American Redstart populations declined at an annual rate of 4.9 percent ($p < 0.1$) (BBS data).

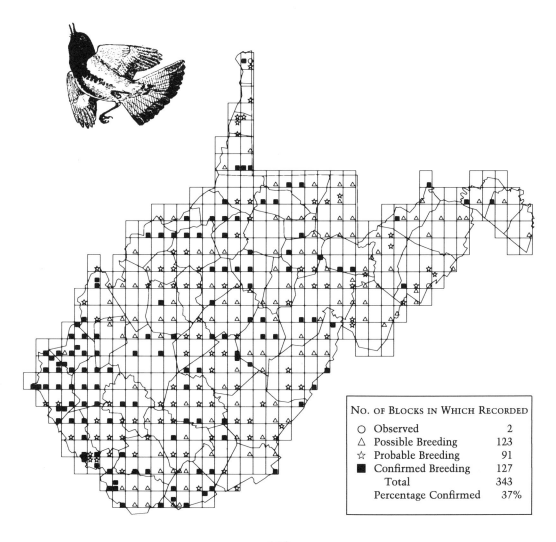

NO. OF BLOCKS IN WHICH RECORDED	
O Observed	2
△ Possible Breeding	123
☆ Probable Breeding	91
■ Confirmed Breeding	127
Total	343
Percentage Confirmed	37%

Prothonotary Warbler *Protonotaria citrea*

The brightly colored PROTHONOTARY WARBLER is a common resident of the southern swamps where its loud song is a characteristic summer sound. It nests in swamps and along large streams from southern Wisconsin, southern Michigan, central New York, and southern New Jersey, south to the Gulf Coast and central Florida. It is largely missing from the highlands of the Appalachians (AOU 1983). Eastern populations showed a 4.4 percent per year increase ($p < 0.01$) from 1966 to 1978, but there has been no significant change since then (Robbins, Sauer, Greenberg, and Droege 1989).

In West Virginia, except for a few scattered records elsewhere, the occurrence is along the lower Ohio River, the Potomac River, and, in particular, the Shenandoah River. This warbler places its nest in a hole in a tree snag, usually over standing water. These nests are easily found with this conspicuous bird, and the percentage of "confirmed" Atlas records is correspondingly high.

The channelization of some streams and the development of shorelines with the elimination of wetlands presents a threat to the Prothonotary Warbler, but the recent formation of the Ohio River Islands National Wildlife Refuge may provide new habitat for this bird.

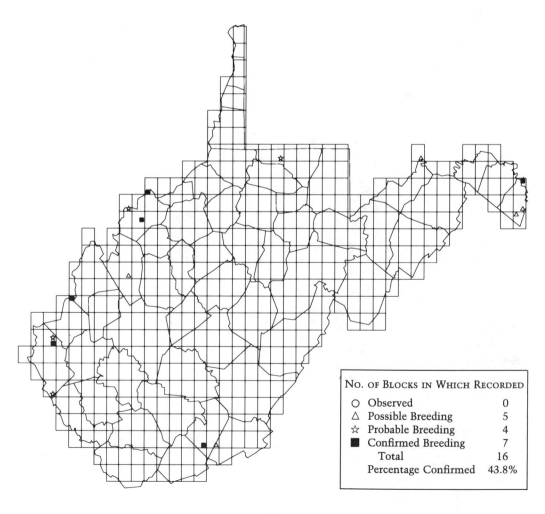

NO. OF BLOCKS IN WHICH RECORDED	
O Observed	0
△ Possible Breeding	5
☆ Probable Breeding	4
■ Confirmed Breeding	7
Total	16
Percentage Confirmed	43.8%

Worm-eating Warbler *Helmitheros vermivorus*

The WORM-EATING WARBLER is one of the least familiar species. Throughout most of its range, it occurs only locally, and it is seldom found in very high densities.

This warbler breeds from southern Ohio, central Pennsylvania, and southern New England, south to southern Mississippi, southern Alabama, central Georgia, western South Carolina, and western Virginia (AOU 1983). Atlas projects in both New York (Bonney 1988) and Pennsylvania (Brauning 1992) found the Worm-eating Warbler at scattered stations north of this indicated range. Eastern populations increased at an annual rate of 1.8 percent ($p < 0.01$) between 1966 and 1978, declining since then, although not at a statistically significant level (Robbins, Sauer, Greenberg, and Droege 1989).

Hall (1983) suspected that the Worm-eating Warbler may nest in every county in West Virginia, but Atlas workers found it only rarely in the Allegheny Mountains Region and in the northern counties. The species' greatest concentration is along the lower Ohio River, where the Ohio Atlas also found it widespread (Peterjohn and Rice 1991), and in the Ridge and Valley Region.

The Worm-eating Warbler is a species of the middle elevations, although occasionally it nests as high as 1,200 meters (Hall 1983). The habitat is mature deciduous woodland that is partly open and lacks dense ground cover. The nests, which are built on the ground, are especially hard to find. The male provides little help to the nest searcher. He often sits on one perch for long periods of time, occasionally delivering his dry, insectlike song, which may be mistaken for that of a Chipping Sparrow.

During the period from 1966 to 1989, the BBS data for West Virginia Worm-eating Warbler populations showed a downward trend. Although not statistically significant, the trend indicates a need for careful monitoring of the numbers of this Neotropical migrant. Because the Worm-eating Warbler seems to require a minimum of 150 hectares, with the smallest plots being 21 hectares (Robbins, Dawson, and Dowell 1989), continued forest fragmentation may cause populations to decrease.

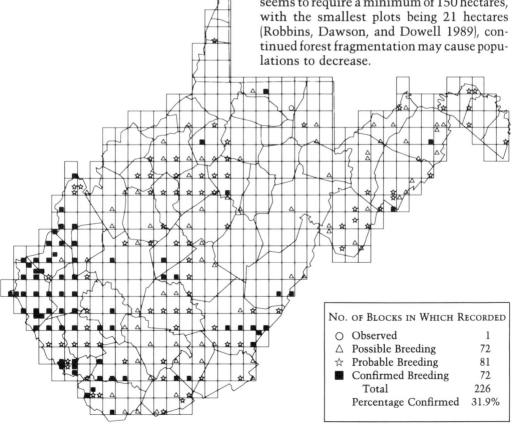

NO. OF BLOCKS IN WHICH RECORDED	
○ Observed	1
△ Possible Breeding	72
☆ Probable Breeding	81
■ Confirmed Breeding	72
Total	226
Percentage Confirmed	31.9%

Swainson's Warbler *Limnothlypis swainsonii*

In the past, the SWAINSON'S WARBLER was thought to be a rather rare and local breeder in the southern states, particularly in cane-brakes and wooded swamps. In the 1930s, however, a population was discovered in the Appalachian Mountains. The range of the species is now known to extend from southern Illinois, western Kentucky, central Tennessee, western West Virginia, and southern Virginia and the Maryland Eastern Shore, south to the Gulf Coast and northern Florida (AOU 1983). In this range, it occurs very locally, often clumped in little pockets.

Hall (1983) listed summer records for 15 West Virginia counties, with nesting records only in Kanawha and Nicholas counties. Atlas volunteers, however, found them in only nine counties, one of which—McDowell—was not on Hall's list. Surprisingly, atlasers did not find this warbler in Kanawha County, which once had the most frequently studied populations; there was only one record in Nicholas County, where W. Legg carried out the pioneering studies on the Appalachian population (Legg 1942).

It is almost certain that the species occurs more widely than the Atlas map indicates, but the localized nature of the species' habitat may not always coincide with the artificial Atlas grid, which samples only one-sixth of the area. Some observers also may not have been familiar with the song of this species, which can be confused with that of the Louisiana Waterthrush. Much of the once-prime habitat, the "hollows" on the south side of the Kanawha River, may have been developed enough to deprive this species of its preferred habitat. An intensive survey of the Swainson's Warbler's known range, much like a previous project by Hurley (1972), would be worthwhile.

The species requires a habitat of dense understory under an older forest, and in many places the understory is supplied by rhododendron thickets. The nest, a bulky cup placed within a meter or two of the ground, is difficult to find in the dense habitat.

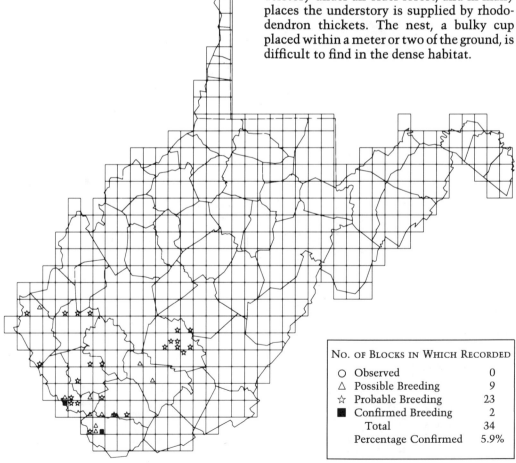

NO. OF BLOCKS IN WHICH RECORDED	
O Observed	0
△ Possible Breeding	9
☆ Probable Breeding	23
■ Confirmed Breeding	2
Total	34
Percentage Confirmed	5.9%

Ovenbird *Seiurus aurocapillus*

The loud *teacher, teacher* song of the OVEN-BIRD is a familiar sound in the eastern deciduous forest, but the bird is a difficult one to see. Its breeding range extends from northern Ontario, northern Quebec, and Newfoundland, south to northern Mississippi, northern Alabama, northern Georgia and eastern North Carolina (AOU 1983). Its favorite breeding habitat is a mature deciduous forest, with little understory but with a heavy ground cover of dead leaves. In the North, the Ovenbird inhabits the mixed forest or aspen woodland. Populations in eastern North America have declined at an annual rate of 1.0 percent ($p < 0.05$) from 1978 to 1987 (Robbins, Bystrak, and Geissler 1986).

Atlas workers found the Ovenbird nesting throughout the state in the mature hard-woods forest at all elevations. Most of the vacant areas on the distribution map either are near centers of population, where woodlands with suitable understory may be missing, or are artifacts of the limited Atlas grid. West Virginia populations showed no significant change between 1966 and 1989 (BBS data), but Hall (1983) commented on a decline in some parts of the state that started much earlier, before the Breeding Bird Survey project started.

The number of "confirmed" breeding records is good for this species, but only 15 of them were made by finding nests. The nest, a depression in the leaf litter with an arched cover of dead leaves, is often found only by accident when the bird flushes from it and then engages in distraction display.

The Ovenbird is a forest-interior species and may be sensitive to habitat fragmentation, but Robbins, Dawson, and Dowell (1989) found it in tracts as small as 1 hectare and listed 6 hectares as the probable minimum size. At these small areas, it will be very susceptible to cowbird parasitism.

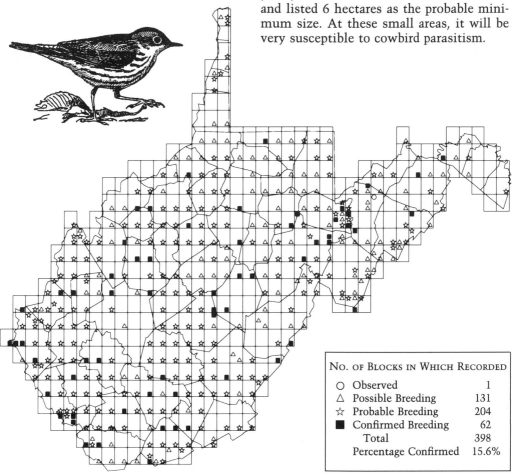

NO. OF BLOCKS IN WHICH RECORDED	
○ Observed	1
△ Possible Breeding	131
☆ Probable Breeding	204
■ Confirmed Breeding	62
Total	398
Percentage Confirmed	15.6%

Northern Waterthrush *Seiurus noveboracensis*

The NORTHERN WATERTHRUSH nests from the northern limit of trees, south to northern Minnesota, central Wisconsin, central Michigan, northeastern Ohio, central Pennsylvania, and northern New Jersey, with a southward extension along the mountains to southeastern West Virginia (AOU 1983). Its favored habitat is at the edge of bogs or near beaver ponds, and it does not use flowing water as consistently as does its relative the Louisiana Waterthrush.

The continental Northern Waterthrush populations increased at the rate of 5.3 percent per year (p <0.01) in the period from 1966 to 1978, but since then there has been no significant trend (Robbins, Sauer, Greenberg, and Droege 1989).

In West Virginia the Northern Waterthrush is limited to the Allegheny Mountains, with most records coming from above 1,000 meters. In some places, the Northern and Louisiana waterthrushes may nest very close together. The BBS data for the Northern Waterthrush indicate an alarming annual rate of decline of 7.95 percent (p <0.01) during the period from 1980 to 1989, but only three routes reported the species.

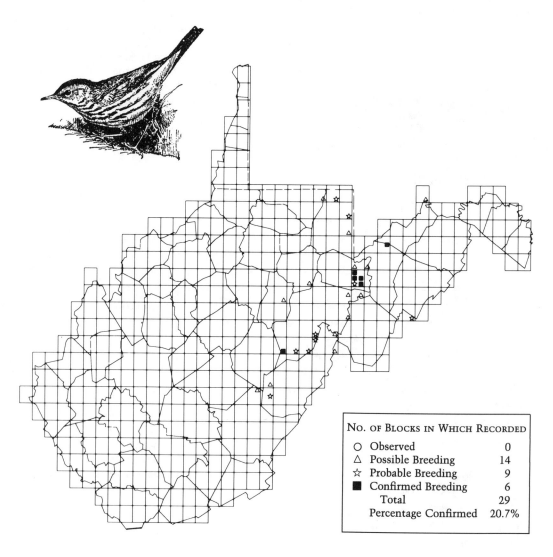

No. of Blocks in Which Recorded	
O Observed	0
△ Possible Breeding	14
☆ Probable Breeding	9
■ Confirmed Breeding	6
Total	29
Percentage Confirmed	20.7%

Louisiana Waterthrush *Seiurus motacilla*

The loud song of the LOUISIANA WATER-THRUSH identifies one of the earliest spring migrant warblers. A southern species, it is common along flowing streams from southern Minnesota, central Wisconsin, southern Michigan, central New York, and southern New England, south to southern Mississippi, southern Alabama, and northern Florida (AOU 1983). It holds a linear territory along streams flowing through heavily wooded valleys. Eastern populations have showed no significant trend since 1966 (Robbins, Sauer, Greenberg, and Droege 1989).

The Louisiana Waterthrush nests throughout West Virginia. It occurs along flowing streams in all types of deciduous forest, even up to altitudes of 1,000 meters. From the Atlas map, it appears to be scarce along the western side of the mountains, which may well be a result of the lack of suitable streams in that area. Other gaps in the map may reflect a similar lack of flowing streams, a lack of coverage in some places, or the bird's requirement for a large habitat area. The Louisiana Waterthrush is an area-sensitive species, and it appears to require an area of 350 hectares to maintain a population. The smallest tracts on which it was found were about 25 hectares (Robbins, Dawson, and Dowell 1989).

West Virginia populations declined at an annual rate of 2.7 percent ($p < 0.10$) in the period from 1966 to 1989 (BBS data).

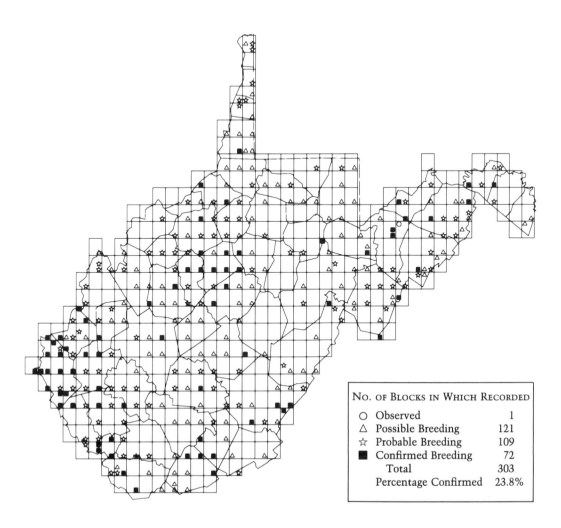

NO. OF BLOCKS IN WHICH RECORDED	
○ Observed	1
△ Possible Breeding	121
☆ Probable Breeding	109
■ Confirmed Breeding	72
Total	303
Percentage Confirmed	23.8%

Kentucky Warbler *Oporornis formosus*

As with other members of the genus *Oporornis*, the KENTUCKY WARBLER usually keeps itself well hidden in its dense habitat. It occurs commonly in the dense understory of mature humid deciduous forests (AOU 1983), but in West Virginia it also occupies the drier oak-pine forest and, occasionally, the northern hardwood forest, even as high as 1,000 meters (Hall 1983). This southern species' range runs from northern Illinois, northern Indiana, northern Ohio, and northern New York, south to the Gulf Coast, southern Georgia, and eastern North Carolina (AOU 1983). Populations in the eastern United States showed no significant trend during the BBS period (Robbins, Sauer, Greenberg, and Droege 1989).

West Virginia Atlas data show that the species is widespread in the Western Hills Region of the state. The few blank places from that area on the Atlas map probably result from lack of coverage. The Kentucky Warbler occurs at a few places in the Allegheny Mountains Region and is uncommon and local in the Ridge and Valley Region. The percentage of "confirmed" records was high (30.4%), and about one-third of these were of nests. West Virginia populations showed no significant trend during the period from 1966 to 1989 (BBS data).

The Kentucky Warbler is a forest-interior species but seems not to be as sensitive to habitat fragmentation as some other species. It was found on tracts as small as 9 hectares and appears to have a minimum viable area of 17 hectares (Robbins, Dawson, and Dowell 1989). Forest fragmentation does, however, increase the amount of cowbird parasitism, and this warbler should be monitored in the future.

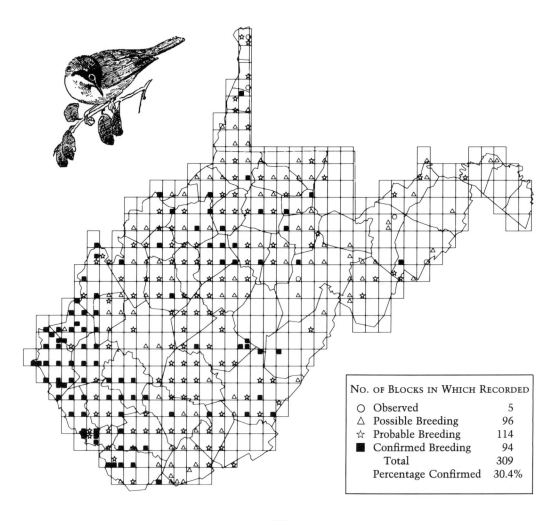

NO. OF BLOCKS IN WHICH RECORDED	
O Observed	5
△ Possible Breeding	96
☆ Probable Breeding	114
■ Confirmed Breeding	94
Total	309
Percentage Confirmed	30.4%

Mourning Warbler *Oporornis philadelphia*

For the average bird watcher, the MOURNING WARBLER is one of the most sought-after species. This warbler's range is limited, and it inhabits the dense, brushy stages of forest succession. It is easy to locate by song but seldom comes out in good view. Unlike some other species, it is able to take advantage of very small patches of habitat, often taking up residence in a small blow-down opening in the mature forest.

Although it is easy to locate where it does occur, it is difficult to see. The Mourning Warbler nests in deciduous areas in the northern forest area from the limit of trees, south to northern Minnesota, northern Wisconsin, southern Michigan, northern Ohio, northern Pennsylvania, and southern New England, with an extension along the high Appalachians south to West Virginia (AOU 1983). Populations in the eastern United States have shown a small (statistically insignificant) decline over the BBS time period (Robbins, Sauer, Greenberg, and Droege 1989).

Hall (1983) reported that the Mourning Warbler nested only in the Allegheny Mountains at elevations above 900 meters, and the Atlas data support this statement. The percentage of "confirmed" Atlas records was high, but only one of these was of finding a nest.

Recent logging operations in the Monongahela National Forest have been beneficial for this species because large numbers of these warblers appear in a clearcut area two or three years after cutting. Such high populations last for five to six years until the new growth becomes too tall for suitable habitat. Because of increased logging operations, the Mourning Warbler population in West Virginia has probably increased in recent years. However, too few BBS routes have reported the species to determine any trend, and no other accurate population data exist.

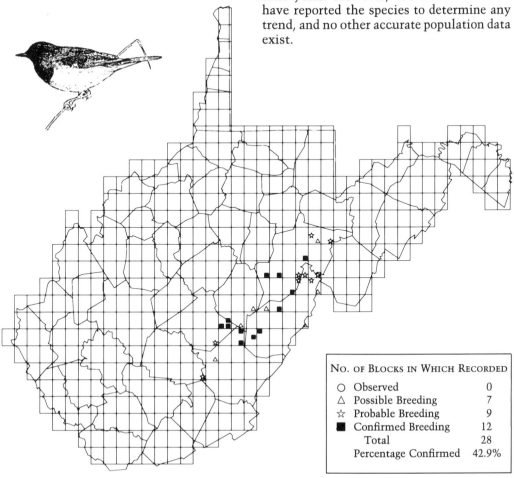

NO. OF BLOCKS IN WHICH RECORDED	
○ Observed	0
△ Possible Breeding	7
☆ Probable Breeding	9
■ Confirmed Breeding	12
Total	28
Percentage Confirmed	42.9%

Common Yellowthroat *Geothlypis trichas*

The COMMON YELLOWTHROAT, with its cheerful *witchity witchity* song, is one of the two most frequently encountered warblers in West Virginia and the eastern United States. The species nests throughout eastern North America south of the tree line, and it is usually numerous where found.

The Yellowthroat's most common habitat is in marshy areas, but it often nests in weedy old fields, in brushy areas along roads and streams, and in brushy second-growth woodland. In West Virginia, where these habitats are quite common, the species occurs at all elevations. Unlike some other species of this habitat, the Yellowthroat is not shy and difficult to see, and so its Lone Ranger-like appearance is well known to birders and nonbirders alike.

Atlas volunteers found the Common Yellowthroat in all parts of the state, and the percentage of "confirmed" records of nesting was high (29.3%). The BBS data, however, show a 2.3 percent per year decline ($p < 0.05$) over the period spanning 1966 to 1989. The decrease in between 1980 and 1989 was 5.3 percent per year ($p < 0.01$).

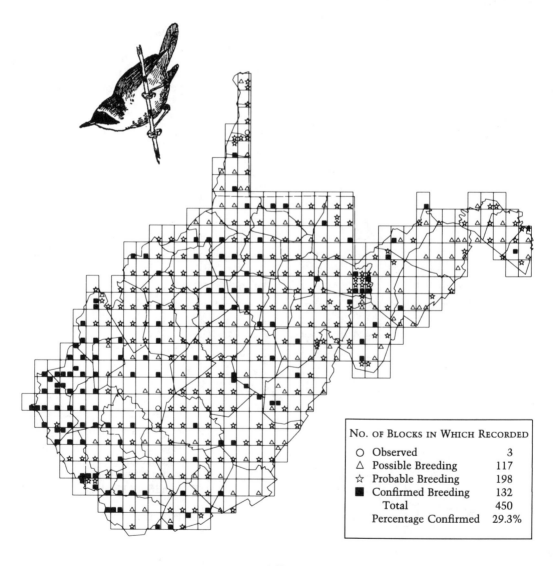

NO. OF BLOCKS IN WHICH RECORDED	
O Observed	3
△ Possible Breeding	117
☆ Probable Breeding	198
■ Confirmed Breeding	132
Total	450
Percentage Confirmed	29.3%

Hooded Warbler *Wilsonia citrina*

For most bird students, the HOODED WARBLER is a loud, clear voice from the tangled undergrowth in a steep ravine. This warbler inhabits the denser tangles, usually in mature deciduous forest, and it is difficult to see without a decided effort.

A southern species, it ranges from central Illinois, southern Michigan, northwestern Pennsylvania, southeastern New York, and southern New England, south to the Gulf Coast and northern Florida (AOU 1983). Eastern populations have remained fairly stable during the years from 1966 to 1989 (Robbins, Sauer, Greenberg, and Droege 1989).

In West Virginia, the Hooded Warbler is common throughout the Western Hills Region, occurring in all forest types. It occurs sparingly in the Allegheny Mountains, often as high as 1,100 meters (Hall 1983). East of the mountains in the Ridge and Valley Region, it becomes local and uncommon.

The Hooded Warbler is another of the forest-interior species that is quite vulnerable to habitat fragmentation and possible cowbird parasitism. Despite these threats, neither the eastern United States nor the West Virginia populations have shown significant change over the period from 1966 to 1989 (BBS data). Nevertheless, the species should continue to be monitored.

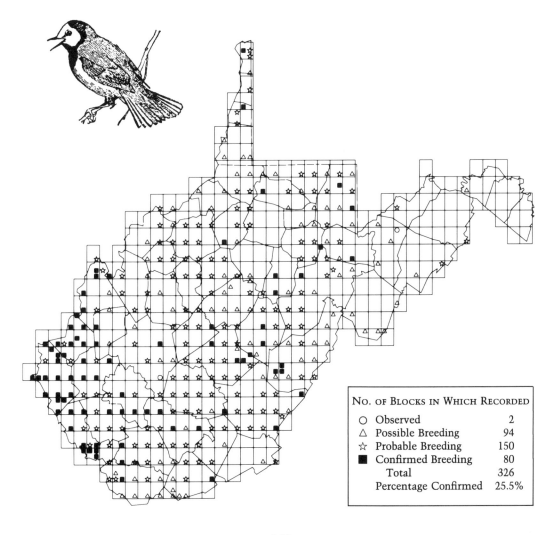

NO. OF BLOCKS IN WHICH RECORDED	
O Observed	2
△ Possible Breeding	94
☆ Probable Breeding	150
■ Confirmed Breeding	80
Total	326
Percentage Confirmed	25.5%

Canada Warbler *Wilsonia canadensis*

The CANADA WARBLER is a northern species whose range extends across the boreal forest, south to northern Wisconsin, northern Michigan, central Pennsylvania, and southern New England, with a projection south along the higher Appalachians to northwestern Georgia (AOU 1983). It inhabits the undergrowth of open woodland and is also found in road cuts and along the edges of ponds and bogs. In West Virginia it also occurs in the undergrowth of the mature spruce-hardwoods forest. Eastern populations have experienced a decline of 2.7 percent per year ($p < 0.01$) (Robbins, Sauer, Greenberg, and Droege 1989).

In West Virginia the Canada Warbler is found in the undergrowth of the northern forests, usually above 650 meters but occasionally lower (Hall 1983). The Atlas records were confined to the Allegheny Mountains Region. Too few BBS routes in West Virginia have reported this species to draw any conclusions on population trends.

Nests of this ground-nesting, thicket-inhabiting species are hard to find, and all of the "confirmed" records were of sightings of young out of the nest.

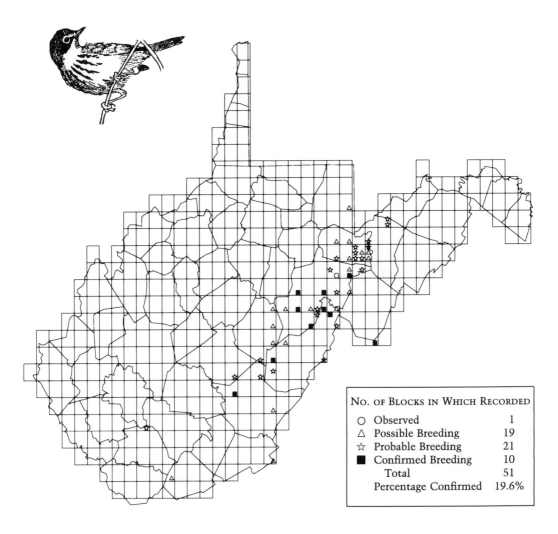

NO. OF BLOCKS IN WHICH RECORDED	
○ Observed	1
△ Possible Breeding	19
☆ Probable Breeding	21
■ Confirmed Breeding	10
Total	51
Percentage Confirmed	19.6%

Yellow-breasted Chat *Icteria virens*

With its raucous song coming loudly from a tangle of blackberry bushes, the YELLOW-BREASTED CHAT is a familiar bird to most West Virginia bird students. Largely a southern species, it nests from southern Iowa, southern Wisconsin, southern Michigan, and southern New York, south to the Gulf Coast and northern Florida (AOU 1983). Populations east of the Mississippi River declined at an annual rate of 4.5 percent ($p < 0.01$) in the years 1966 to 1978, but they showed an increase of 1.2 percent per year ($p < 0.01$) from 1978 to 1987 (Robbins, Sauer, Greenberg, and Droege 1989).

The chat nests in the brushy stages of plant succession in West Virginia, espe-cially in fields overgrown with *Rubus* species, where the nest is often built in the densest cover available. The bird is most common at low elevations, but it does occur in suitable habitat on some of West Virginia's higher peaks.

Atlas workers found the Yellow-Breasted Chat to be distributed widely except in the Allegheny Mountains. The West Virginia population has undergone a decline at the rate of 5 percent per year ($p < 0.01$) during the period from 1966 to 1989, although the decline was negligible in the decade from 1980 to 1989 (BBS data). Some of this decline resulted when habitat outgrew the suitable stage, but many areas of apparently still-suitable habitat are without chats.

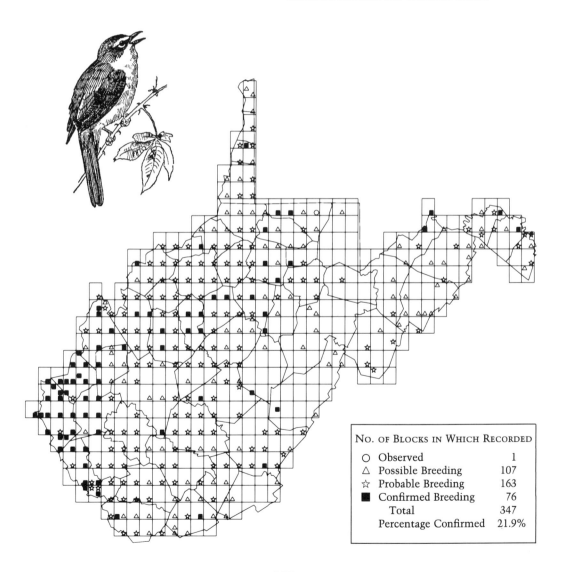

NO. OF BLOCKS IN WHICH RECORDED	
○ Observed	1
△ Possible Breeding	107
☆ Probable Breeding	163
■ Confirmed Breeding	76
Total	347
Percentage Confirmed	21.9%

Summer Tanager *Piranga rubra*

As with its more familiar relative, the Scarlet Tanager, the SUMMER TANAGER is more often heard than seen. It inhabits the canopies of mature forest and is most common in the oak-hickory and oak-pine forests. It occurs in more open situations than does the Scarlet Tanager.

The range of this southern bird extends from southern Iowa, southern Illinois, southern Indiana, southern Ohio, southwestern Pennsylvania, and southern New Jersey, south to southern Florida (AOU 1983). The Summer Tanager is highly area-sensitive and seems to require a minimum area of 40 hectares for breeding. It was not located on plots of less than 47 hectares (Robbins, Dawson, and Dowell 1989).

In West Virginia, Atlas workers found the headquarters of the species located in the southwestern part of the state and around the southern Ohio River and its tributaries.

Except for a single occurrence, the Summer Tanager was missing north of Wetzel County in the Ohio Valley and was not found in the Allegheny Mountains or the Monongahela drainage. The Ohio and Pennsylvania Atlas projects (Peterjohn and Rice 1991; Brauning 1992) recorded the Summer Tanager at locations contiguous with the Northern Panhandle (where the West Virginia Atlas had only one record) and Monongalia County (where atlasers did record it sparingly). In the past there has been a substantial population in the Shenandoah and Potomac valleys (Hall 1983). It is difficult to know whether the failure of the Atlas survey to find the Summer Tanager there is an artifact of coverage or a real decline in the population in those areas. Overall, West Virginia populations have shown no significant change in the period from 1966 to 1989 (BBS data).

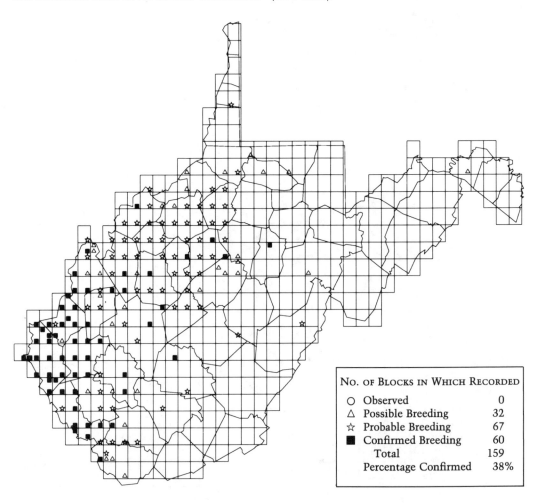

NO. OF BLOCKS IN WHICH RECORDED	
O Observed	0
△ Possible Breeding	32
☆ Probable Breeding	67
■ Confirmed Breeding	60
Total	159
Percentage Confirmed	38%

Scarlet Tanager *Piranga olivacea*

Despite its colorful plumage, the SCARLET TANAGER is a difficult bird to see and is more often heard than seen. It inhabits the leafy canopies of mature forests and seldom comes into plain view. It nests in mature deciduous forest from southern Ontario, Quebec, New Brunswick, and central Maine, south to northern Alabama and northern Georgia (AOU 1983). Populations in the eastern United States declined at a rate of 1.2 percent per year (p <0.01) between 1978 and 1987 (Robbins, Sauer, Greenberg, and Droege 1989).

Atlas workers found the Scarlet Tanager throughout West Virginia. The largest populations are in the oak-hickory forest, but it also occurred in the northern hardwoods and the mixed hardwoods-spruce forest at high elevations. Between 1966 and 1989, the population in the state has increased at an annual rate of 2.6 percent (p <0.01) (BBS data).

The Scarlet Tanager is very sensitive to area of forest, and the maximum probability of finding the species occurs in forest tracts of greater than 3,000 hectares. Twelve hectares is the suggested minimum area for breeding, although it was found occasionally on tracts as small as 2 to 5 hectares (Robbins, Dawson, and Dowell 1989). The continued well-being of this species thus depends on the presence of sizable tracts of mature forest. The Scarlet Tanager is also vulnerable to habitat loss on its tropical wintering grounds.

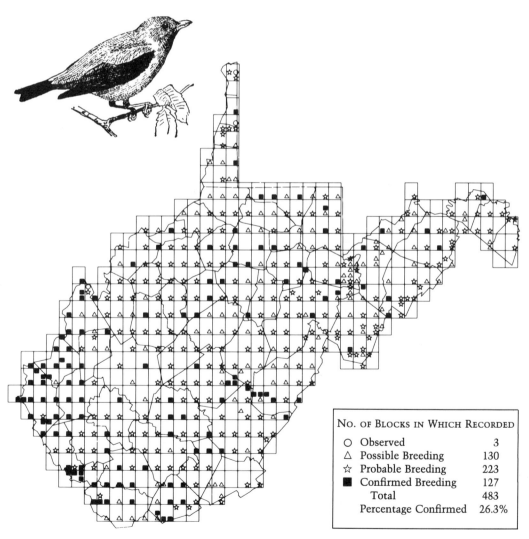

NO. OF BLOCKS IN WHICH RECORDED	
O Observed	3
△ Possible Breeding	130
☆ Probable Breeding	223
■ Confirmed Breeding	127
Total	483
Percentage Confirmed	26.3%

Northern Cardinal *Cardinalis cardinalis*

Few birds are more familiar to West Virginians than the NORTHERN CARDINAL, which is the official state bird. Although originally a southern species, it now nests throughout the eastern United States, south of northern Wisconsin, northern Michigan, and northern New England (AOU 1983).

A year-round resident, the cardinal's bright plumage adds a welcome spot of color to the drab winter scene. It readily comes to feeding stations and it is a common backyard bird through winter. Three or more broods are raised in the breeding season, and it is not uncommon to see adults feeding fledged young in the early fall. Thus a high proportion (52.3%) of the Atlas records were "confirmed."

Atlas workers found the cardinal in all sections of West Virginia. In the mountain region, it was found only in the valleys, but it may occur at elevations as high as 1,000 meters. It nests in forest edge situations, small woodlots, and in suburban gardens. The cardinal usually places its nest, a loosely assembled cup, below 3 meters from the ground in dense shrubbery or a low tree. Populations have remained stable in the state during the period 1966 to 1989 (BBS data).

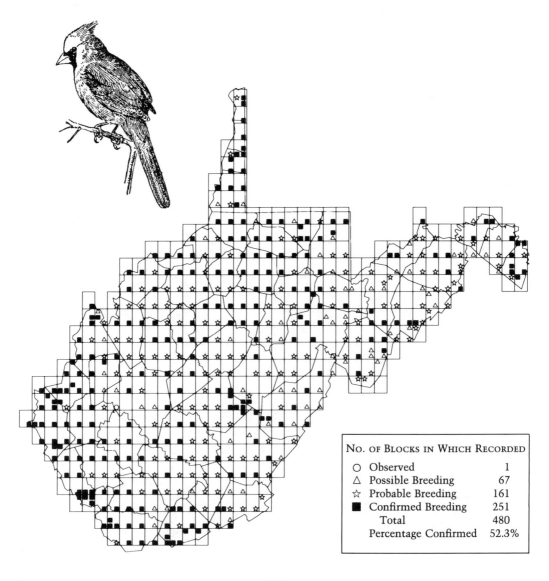

No. of Blocks in Which Recorded	
○ Observed	1
△ Possible Breeding	67
☆ Probable Breeding	161
■ Confirmed Breeding	251
Total	480
Percentage Confirmed	52.3%

Rose-breasted Grosbeak *Pheucticus ludovicianus*

The ROSE-BREASTED GROSBEAK is a bird of the northern deciduous forest. It nests from southern Ontario and southern New Brunswick, south to southern Illinois, central Indiana and Ohio, southern Pennsylvania, and northern New Jersey, with an extension along the Appalachian highlands south to northern Georgia (AOU 1983). After an annual increase of 6.1 percent ($p < 0.01$) from 1966 to 1978, the continental populations declined at a rate of 4.1 percent per year ($p < 0.01$) from 1977 to 1987 (Robbins, Sauer, Greenberg, and Droege 1989).

West Virginia occurrences were concentrated along the highlands of the Allegheny Mountains Region and the higher ridges of the Ridge and Valley Region. The Rose-breasted Grosbeak's range extends into the higher hills of the southern part of the Western Hills. It also nests on the ridges of the Northern Panhandle, although these are lower than the mountains. Its main habitat is the northern hardwoods forest, although it also occurs in the spruce-hardwood forest, as well as the oak forest below the northern hardwoods.

Rose-breasted Grosbeaks enter wooded tracts in the middle stages of successional growth and remain until the climax is reached, but maximum numbers occur before the forest has reached the fully mature stage. The nest is a rather flimsy affair placed in a fork, usually in a low tree. Populations in the state showed an annual increase of 4.7 percent ($p < 0.05$) in the period spanning 1966 to 1989 (BBS data).

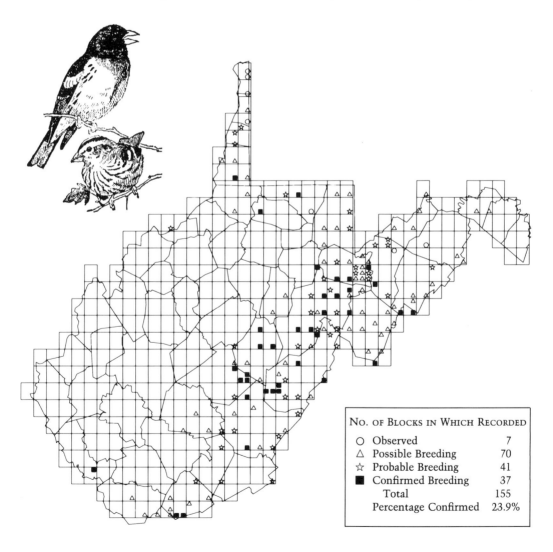

NO. OF BLOCKS IN WHICH RECORDED	
○ Observed	7
△ Possible Breeding	70
☆ Probable Breeding	41
■ Confirmed Breeding	37
Total	155
Percentage Confirmed	23.9%

Blue Grosbeak *Guiraca caerulea*

The BLUE GROSBEAK is a bird of the southern and western United States. Its eastern range extends as far north as southwestern Iowa, northern Missouri, central Illinois, Indiana, and Ohio, through southern and eastern West Virginia to southeastern Pennsylvania (AOU 1983). Nesting occurs in open areas with scattered trees, scrub, thickets, and cultivated land. The species has been actively extending its range into West Virginia, Ohio, and southern Pennsylvania. Breeding Bird Survey data from 1978 to 1987 showed an annual rate of increase of 2.7 percent ($p < .05$) (Robbins, Sauer, Greenberg, and Droege 1989).

Until recently, the Blue Grosbeak was known in West Virginia only in the broad Shenandoah and South Branch valleys in the Ridge and Valley Region. Its favorite habitat was the orchards so common in that area (Hall 1983). During the Atlas project, volunteers found numerous records in this region. In recent years, this grosbeak has entered the lower Ohio Valley and the valleys of some of the main tributaries of the Ohio. Its habitat there is often open land overgrown with *Crataegus* or other shrubs. There are a few scattered records from the central part of the state.

Blue Grosbeaks are easy to find, and the vacant areas in the Atlas map do not seem to be artifacts of coverage. BBS data show that the species has increased at an annual rate of 7.1 percent from 1966 to 1989, but the number of routes is too small to evaluate the significance of this change. Further expansion of the range, with increasing numbers, can be expected in the future.

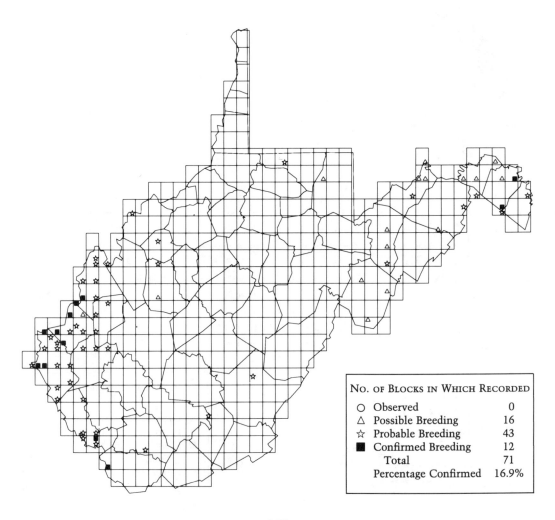

No. of Blocks in Which Recorded	
○ Observed	0
△ Possible Breeding	16
☆ Probable Breeding	43
■ Confirmed Breeding	12
Total	71
Percentage Confirmed	16.9%

Indigo Bunting *Passerina cyanea*

The INDIGO BUNTING is a familiar bird to most Easterners. The Bunting's breeding range extends from northern Minnesota, southern Ontario, southwestern Quebec, and southern Maine, south to the Gulf Coast (AOU 1983). Nesting occurs in clearings, open second-growth forests, and particularly in forest edge situations, such as road clearings. The nest is usually placed close to the ground in dense plant cover. Populations in the eastern states showed an annual decline of 0.7 percent from 1978 to 1987 ($p < 0.01$) (Robbins, Sauer, Greenberg, and Droege 1989).

Because of its attachment to edge and roadside situations, the Indigo Bunting is perhaps the most common bird observed as one drives through wooded West Virginia in the summer. Atlas workers found it in all parts of the state and at all elevations. It was the third most widely reported species, and the number of "confirmed" records of nesting is high (40.6%). Breeding Bird Survey data from 1966 to 1989 showed an annual decline of 1.0 percent ($p < 0.05$).

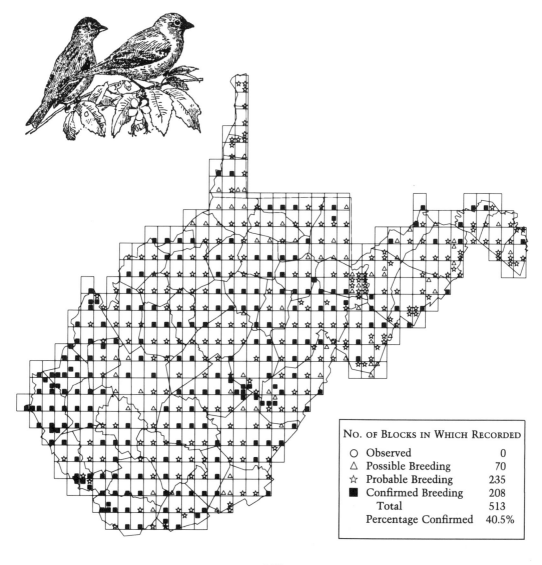

No. of Blocks in Which Recorded	
○ Observed	0
△ Possible Breeding	70
☆ Probable Breeding	235
■ Confirmed Breeding	208
Total	513
Percentage Confirmed	40.5%

Dickcissel *Spiza americana*

The DICKCISSEL, a bird of the midwestern grasslands, normally breeds as far east as western Ohio, Kentucky, and Tennessee. In the past, it has nested farther east, as it does sporadically today (AOU 1983). From 1965 to 1977, populations in the eastern part of the range dropped significantly, although the species is known for major fluctuations in population and range from year to year (Robbins, Bystrak, and Geissler 1986).

The Dickcissel has been reported in West Virginia from time to time and has probably nested there in the past, although Hall (1983) had no definite nesting records. In 1988, when a severe drought occurred in this species' main range, Dickcissels appeared in many places east of their normal distribution. Nestings were established in south-western Pennsylvania and Virginia (Hall 1988). During that year, West Virginia sightings came from Jefferson, Hardy, Grant, Wood, and Wirt counties. The "confirmed" nestings in Grant, Hardy, and Wirt counties were the first documented nestings for the state. The nestings were in fields covered with tall grasses or mixtures of grasses and forbs. The Grant County birds were not present in 1989, and the situation in the other counties is not known. There had been one earlier Atlas record from Mason County.

The Dickcissel is thought to be at some hazard in its main range because the species' sex ratio is greatly unbalanced in favor of males. It is also subject to poisoning by agricultural chemicals used in Middle America, where it winters.

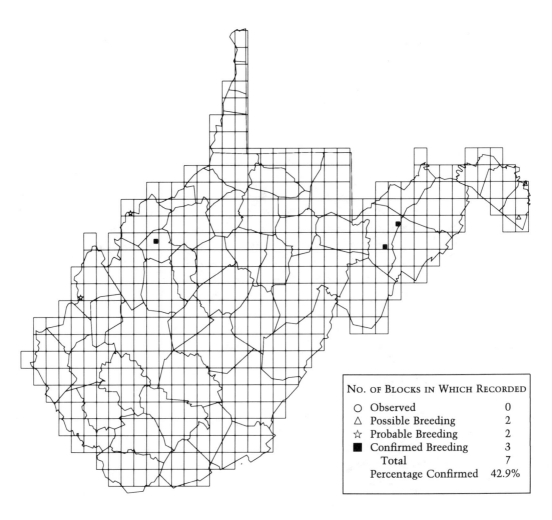

NO. OF BLOCKS IN WHICH RECORDED	
O Observed	0
△ Possible Breeding	2
☆ Probable Breeding	2
■ Confirmed Breeding	3
Total	7
Percentage Confirmed	42.9%

Rufous-sided Towhee *Pipilo erythrophthalmus*

The Rufous-sided Towhee is a common bird, nesting throughout the eastern United States south of northern Minnesota and northern New England. It nests in open woodland, forest edge, second growth, brushy areas, and riparian thickets (AOU 1983). Eastern populations are declining slightly (Robbins, Bystrak, and Geissler 1986).

The Atlas project found the towhee nesting throughout West Virginia in suitable habitat and at all elevations. It tied with the American Robin as the species reported from the most blocks in the Atlas survey. The number of "confirmed" records was high (31.5%).

BBS data, however, show that the Rufous-sided Towhee has undergone a decline at the rate of 2.5 percent per year ($p < 0.01$). This represents a 44 percent decline in the 23 years of the survey period. Some of this decline can be attributed to the disappearance of suitable habitat.

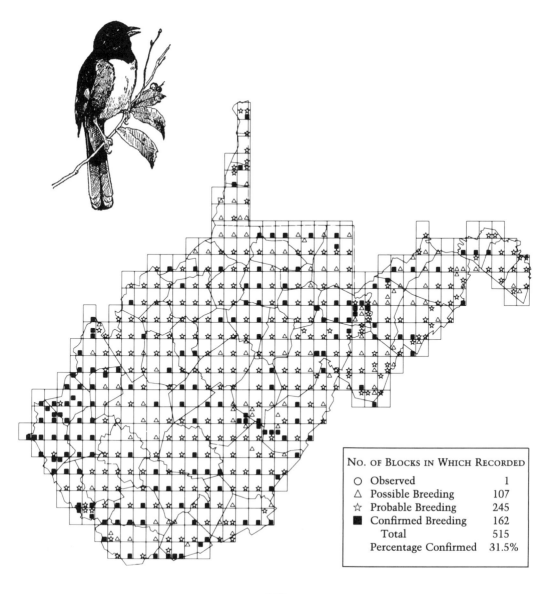

No. of Blocks in Which Recorded	
○ Observed	1
△ Possible Breeding	107
☆ Probable Breeding	245
■ Confirmed Breeding	162
Total	515
Percentage Confirmed	31.5%

Bachman's Sparrow *Aimophila aestivalis*

The BACHMAN'S SPARROW was originally a bird of the open pine woods of the southern United States. In the early part of this century, however, it spread northward, inhabiting brushy overgrown fields, and its range extended as far north as central Indiana and Ohio and southwestern Pennsylvania. The history of this expansion was reported by Brooks (1938a). Starting in the 1940s, the species began to disappear from the northern part of its range, and it is now absent or local north of southern Kentucky and North Carolina (AOU 1983).

The species had once been widespread and moderately common in all of West Virginia west of the Allegheny Mountains (Brooks 1938a). In this range, it inhabited abandoned pastures that were beginning to grow up in shrubs. The Bachman's Sparrow has now nearly totally disappeared in the state, and it has been designated a species of special concern in West Virginia (W. Va. DNR n.d.).

There seems to be no evident explanation for the disappearance of the Bachman's Sparrow, either from West Virginia or from the nearby states. Much of its favored habitat has grown to a stage that is not used by the sparrow, but much apparently suitable habitat remains. A recent study in the South (Dunning and Watts 1990) suggested that this sparrow's habitat requirements are very strict, and indeed habitat loss may be the major factor involved in the species' declining numbers. No detailed studies were ever made on the Bachman's Sparrow in West Virginia. In the absence of any knowledge of its requirements in the state, there appears to be no way to improve the status of this species in West Virginia.

There was only one Atlas report. A singing male was seen on several occasions in Monongalia County in 1989.

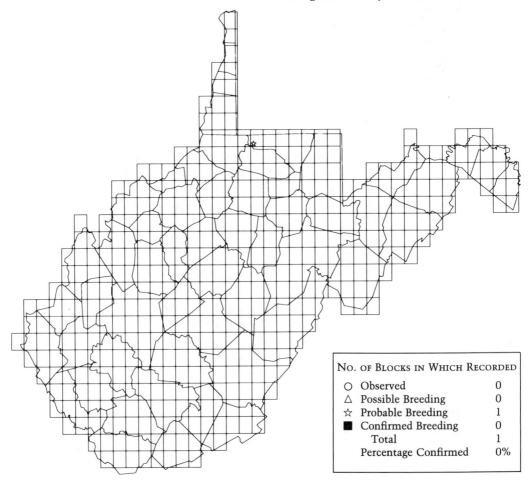

NO. OF BLOCKS IN WHICH RECORDED	
○ Observed	0
△ Possible Breeding	0
☆ Probable Breeding	1
■ Confirmed Breeding	0
Total	1
Percentage Confirmed	0%

Chipping Sparrow *Spizella passerina*

With a range that includes almost all of the United States and parts of southern Canada, the CHIPPING SPARROW is not only one of the most widespread species, but its tame and confiding nature makes it one of the most familiar of the sparrows. In much of its range it is a common dooryard bird, and the dry chipping song from which it takes its name is a familiar sound.

This sparrow nests in open woodlands and forest edge of both deciduous forest and pine-oak formations. It is quite common in suburban gardens, parks, and cemeteries, where it often builds its nest in ornamental conifers.

The Atlas data show this species occurring throughout West Virginia, and the high percentage (64.8%) of "confirmed" records of breeding attests to both the ease of finding the nests and the large populations near human habitation.

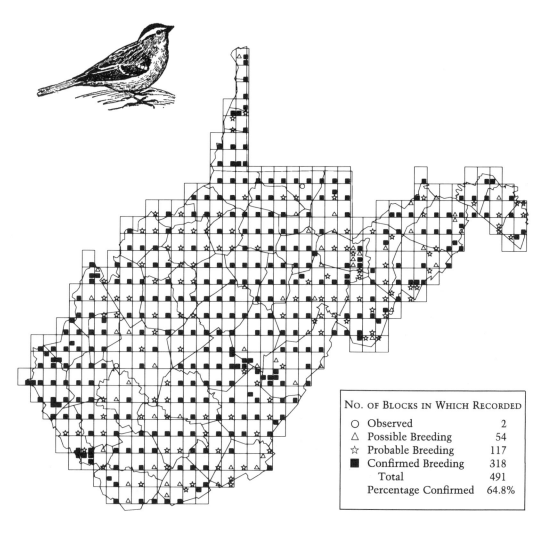

No. of Blocks in Which Recorded	
○ Observed	2
△ Possible Breeding	54
☆ Probable Breeding	117
■ Confirmed Breeding	318
Total	491
Percentage Confirmed	64.8%

Field Sparrow *Spizella pusilla*

The FIELD SPARROW was formerly a common bird of the old fields, brushy hillsides, and forest edge from northern Minnesota, northern Wisconsin, and northern New England, south to the Gulf Coast (AOU 1983). Populations in the East have been declining significantly since 1966, probably because of decreases in suitable habitat (Robbins, Bystrak, and Geissler 1986).

The Field Sparrow breeds in all parts of West Virginia. The gaps in the distribution illustrated in the Atlas map are to some extent artifacts of lack of coverage, but they also reflect decreased habitat in the steep valleys of the southern hills. Although quite abundant until recently, this species is now much reduced. Several hard winters in the late 1970s contributed to this decline, but the lack of suitable brushy fields was the major factor in the species' decline. BBS data showed an annual decline of 4.5 percent ($p < 0.01$) in the period from 1966 to 1989.

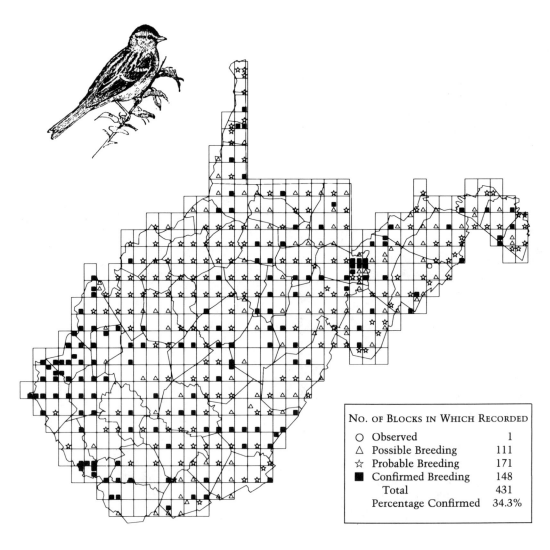

NO. OF BLOCKS IN WHICH RECORDED	
○ Observed	1
△ Possible Breeding	111
☆ Probable Breeding	171
■ Confirmed Breeding	148
Total	431
Percentage Confirmed	34.3%

Vesper Sparrow *Pooecetes gramineus*

The VESPER SPARROW breeds from Northern Manitoba, Ontario, and Maine, south to central Kentucky, North Carolina, and Maryland (AOU 1983). Within this range, it inhabits grasslands and savannas with low trees or scrub. Continental populations have been declining since the beginning of the USFWS Breeding Bird Survey (Robbins, Bystrak, and Geissler 1986).

The distribution of the Vesper Sparrow in West Virginia as determined by the Atlas project differs somewhat from that reported by Hall (1983). Atlas records show that this sparrow was found only sparingly in the center of the state. It seems likely that most of the open fields that have suitable habitat are too small to hold a population of Vesper Sparrows. The largest number of reports came from the Ridge and Valley Region,

where the bird occurs not only in the valleys but also on some of the grassy summits. Although Hall (1983) reported good numbers in the Northern Panhandle, they have disappeared from that area in the last decade, and Atlas workers failed to locate them.

West Virginia populations have been declining at an annual rate of 8.9 percent ($p < 0.01$) during the period spanning 1966 to 1989 (BBS data). This corresponds to a overall decline of 88 percent in the 23-year period. Loss of suitable habitat is the most probable cause of this decline. Some Atlas workers may have been unfamiliar with the song of this species, which may also account for some absences. The Vesper Sparrow should perhaps be added to the list of species of special concern in West Virginia.

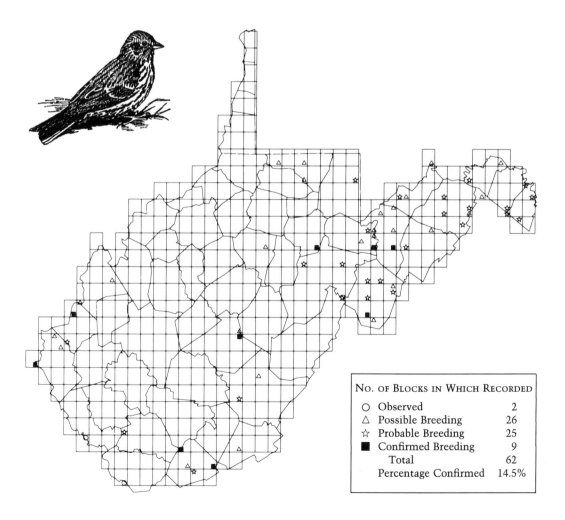

No. of Blocks in Which Recorded	
O Observed	2
△ Possible Breeding	26
☆ Probable Breeding	25
■ Confirmed Breeding	9
Total	62
Percentage Confirmed	14.5%

Lark Sparrow *Chondestes grammacus*

At present, the LARK SPARROW is a breeding bird of the midwestern prairie country and the farmland west of the Mississippi River. It is very local as a breeding bird east of that river, and it has withdrawn from much of the published eastern range (Peterson 1980; AOU 1983). In the latter part of the nineteenth century, the Lark Sparrow extended its range eastward into Pennsylvania, West Virginia, and Virginia. Formerly the range extended much farther east. By the 1920s, the range included most of West Virginia west of the mountains, except for the immediate Ohio River Valley (Hall 1983; Brooks 1938b).

Starting in the 1930s, the Lark Sparrow began to withdraw from the eastern part of its range, and it has now almost disappeared from West Virginia. Hall (1983) reported that the Lark Sparrow was to be found only in two widely separated areas: in the South Branch and North Fork valleys in Pendleton and Hampshire counties; and in Mason and adjoining counties. A nest had been found in Hampshire County in 1982, before the start of the Atlas project (Laitsch 1983). Atlas workers found Lark Sparrows only in Jackson and Putnam counties.

The Lark Sparrow is listed as a species of special concern by the Division of Natural Resources (W. Va. DNR n.d.). No reason is readily apparent for the decline of this sparrow.

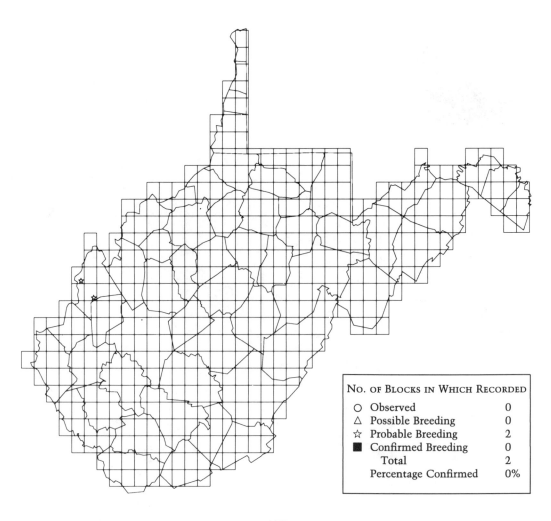

NO. OF BLOCKS IN WHICH RECORDED	
O Observed	0
△ Possible Breeding	0
☆ Probable Breeding	2
■ Confirmed Breeding	0
Total	2
Percentage Confirmed	0%

Savannah Sparrow *Passerculus sandwichensis*

The SAVANNAH SPARROW nests throughout northeastern North America, south to northern Missouri, southern Illinois, central Indiana, central Ohio, northern West Virginia, and southeastern Pennsylvania (AOU 1983). It inhabits open grasslands, marshes, and farmlands. Populations were essentially stable during the period from 1966 to 1979 (Robbins, Bystrak, and Geissler 1986).

In West Virginia, Atlas volunteers found the Savannah Sparrow principally along the axis of the Allegheny Mountains and in the intermountain valleys. Another population was found in the Northern Panhandle and in the southern highlands, areas in which this sparrow is especially common on recovered surface mines. This distribution is almost exactly the same as that mapped by Hall (1983).

The BBS data for West Virginia from 1966 to 1989 show an annual decline of 3.2 percent ($p < 0.01$). Since 1980, the decline has been even greater, at 7.7 percent per year ($p < 0.01$). The reason for this decline is not apparent. While much attention has been focused primarily on the forest birds that winter in the Neotropics, it has become apparent that some of the birds of open non-forest areas that winter in the southern United States are also declining in numbers. The Savannah Sparrow and the Field Sparrow are examples of this.

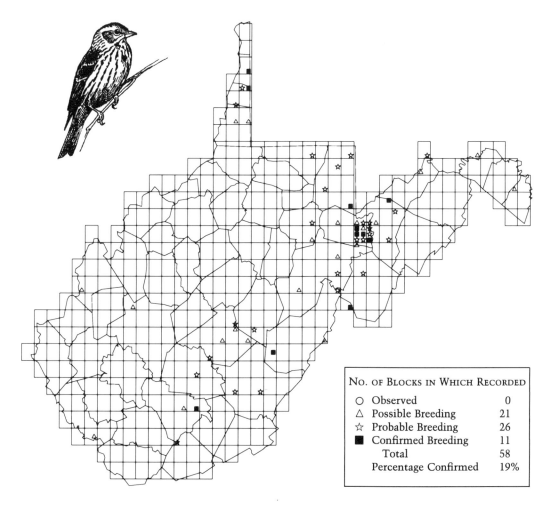

NO. OF BLOCKS IN WHICH RECORDED	
○ Observed	0
△ Possible Breeding	21
☆ Probable Breeding	26
■ Confirmed Breeding	11
Total	58
Percentage Confirmed	19%

Grasshopper Sparrow *Ammodramus savannarum*

A weak, insectlike call coming from the tall grass meadows is the first indication of the presence of the GRASSHOPPER SPARROW for most bird watchers. However, its habit of singing from the top of a plant stalk usually makes it easy to see.

The species' eastern range extends from northern Minnesota and Wisconsin, southern Ontario, and southern New England, south to central Mississippi, central Alabama, central Georgia, and southeastern Virginia (AOU 1983). Its usual habitat is hayfields and other grassy open lands. Populations have declined significantly in the period from 1966 to 1979 (Robbins, Bystrak, and Geissler 1986), and the species has been included on the Audubon Society's Blue List (Tate 1986).

The Atlas project found most records of breeding Grasshopper Sparrows in the East-ern Panhandle, the valleys of the Ridge and Valley Region, the Ohio Valley, and the higher southern part of the Western Hills Region. This sparrow is very common in recovered strip mines in northern West Virginia (Whitmore and Hall 1978). Scattered records were reported from the central part of the state.

This sparrow's absence from many places on the Atlas map is not readily explainable. Although suitable habitat is scarce in the central part of the state, it is not totally absent there. Some observers may have been unable to hear the high-pitched song of this species, and much of the reclaimed strip-mine habitat may have been inaccessible to Atlas workers.

West Virginia's Grasshopper Sparrow population declined by 7.3 percent per year (p <0.01) between 1966 and 1989, according to BBS data.

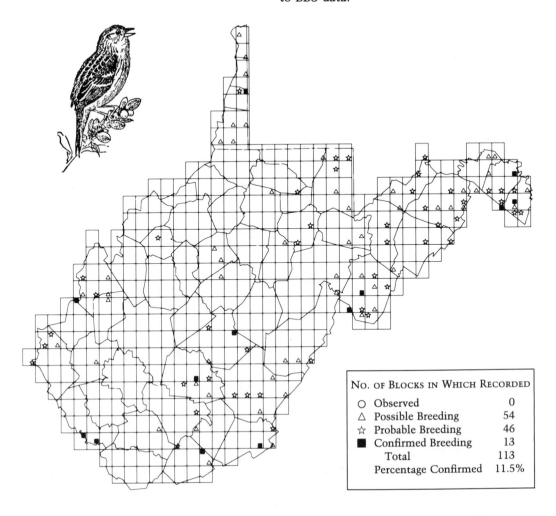

No. of Blocks in Which Recorded	
○ Observed	0
△ Possible Breeding	54
☆ Probable Breeding	46
■ Confirmed Breeding	13
Total	113
Percentage Confirmed	11.5%

Henslow's Sparrow *Ammodramus henslowii*

The HENSLOW'S SPARROW is a secretive, seldom-seen bird of the grassy, weed-filled fields. Its range extends from central Minnesota, central Wisconsin, central Michigan, southern Ontario, northern New York, and southern New England, south to central Kansas, Missouri, southern Illinois, northern Kentucky, central West Virginia, and eastern Virginia (AOU 1983). In the northeastern part of this range, the populations have decreased in recent years (AOU 1983), and this sparrow is almost gone from New England (Peterson 1980). The normal habitat of the Henslow's Sparrow is that of the early years of plant succession, when the fields are growing up to broomsedge (*Andropogon*) and weeds. This habitat is very short-lived and the birds are soon forced to move elsewhere.

In West Virginia, the older fields that this species once used have grown up and no longer provide suitable support for this sparrow. Very few new fields are allowed to stand idle for the few years needed to develop habitat for the Henslow's Sparrow. Although Hall (1983) listed summer records from 24 West Virginia counties, including most of the eastern part of the state, the Atlas survey found it in only seven counties, only one of which was listed by Hall. The Atlas survey undoubtedly missed some records. This species is often not found by roadside birding, and some atlasers may not have been able to hear the Henslow's Sparrow's weak song.

Breeding Bird Survey data showed a 3.9 percent per year decrease (*p* <0.01) during the years 1966 to 1989. The species has been included in *Vertebrate Species of Concern in West Virginia* (W. Va. DNR n.d.).

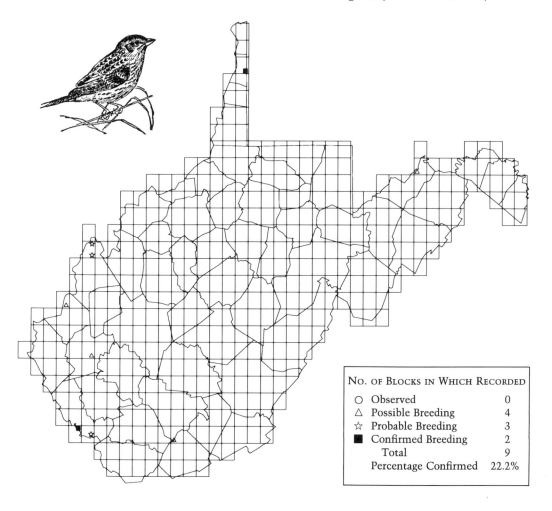

No. of Blocks in Which Recorded	
○ Observed	0
△ Possible Breeding	4
☆ Probable Breeding	3
■ Confirmed Breeding	2
Total	9
Percentage Confirmed	22.2%

Song Sparrow *Melospiza melodia*

The Song Sparrow is a common breeding bird throughout eastern North America (AOU 1983) and one of the most familiar dooryard birds in most of West Virginia. It nests in a variety of places, along forest edges, brushy pastures, and suburban gardens. The Song Sparrow's nest is well hidden in a small tree, shrub, or on the ground beneath a tuft of grass or forbs. It is constructed of grass, stems, or bark fibers and lined with finer plant materials and hair. Three broods are commonly produced in a season. Thus this sparrow can be "confirmed" at any time from spring to early fall.

Atlas workers found the Song Sparrow throughout the state and attained a high percentage of "confirmed" records (48.7%). It was the fifth most widespread species in the state. Only those blocks on which little survey time was spent have less than "probable" status.

Eastern populations showed a significant decline over the years from 1966 to 1979 (Robbins, Bystrak, and Geissler 1986). The Song Sparrow is a frequent host of the Brown-headed Cowbird, which may be contributing to the evident decline of this sparrow.

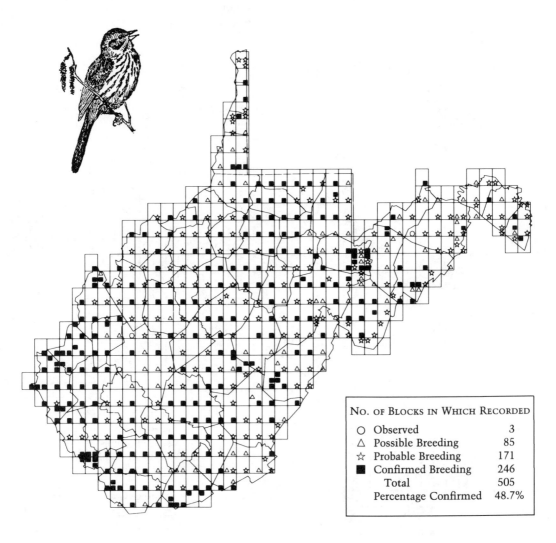

NO. OF BLOCKS IN WHICH RECORDED	
○ Observed	3
△ Possible Breeding	85
☆ Probable Breeding	171
■ Confirmed Breeding	246
Total	505
Percentage Confirmed	48.7%

Swamp Sparrow *Melospiza georgiana*

In eastern North America, the SWAMP SPAR-ROW nests from the limit of trees, south to northern Illinois, northern Indiana, central Ohio, southeastern West Virginia, Maryland, and Delaware (AOU 1983). Nesting is limited to wetlands, bogs, marshes, and wet meadows, all with at least a few low bushes or trees. Because of this limitation, populations generally track the presence of wetlands, and thus they have shown a slight decrease in the eastern United States (Robbins, Bystrak, and Geissler 1986).

The main range in West Virginia, as shown on the Atlas map, is in the high wetlands of the Allegheny Mountains, with heavy concentrations in such places as Ca-

naan Valley. There are a few outlying records. Hall (1983) reported nestings in the past in the Ohio River counties, Hancock, Tyler, Wood, and Pleasants, where the Atlas workers failed to find them. Since it is easy both to observe and to evaluate the breeding status of this bird, one can assume that these absences are due to the present lack of suitable wetland habitat in those areas. The usual nesting habitat of this species is in alder swamps, often connected with beaver ponds, where the nest is usually on a tussock of grass or a low bush.

The future status of the Swamp Sparrow in West Virginia will depend upon the future of the state's wetlands.

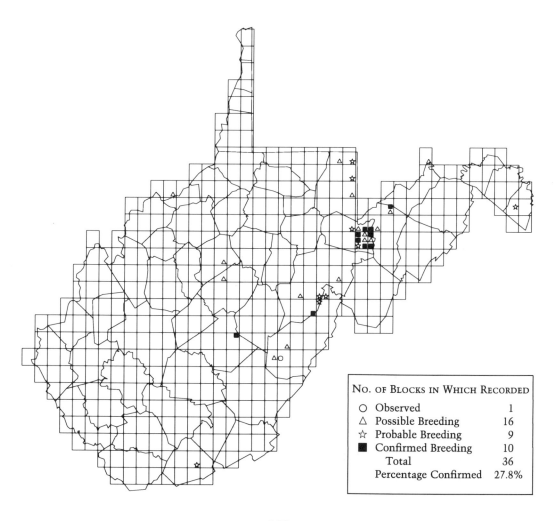

NO. OF BLOCKS IN WHICH RECORDED	
O Observed	1
△ Possible Breeding	16
☆ Probable Breeding	9
■ Confirmed Breeding	10
Total	36
Percentage Confirmed	27.8%

White-throated Sparrow *Zonotrichia albicollis*

The WHITE-THROATED SPARROW is a common bird of the northern forests, both the pure coniferous and the mixed coniferous-hardwoods. It usually occurs in a forest edge situation, in bogs, or in brush thickets. Its range extends from the limit of trees in the North, south to northern Wisconsin, central Michigan, northern Ohio, northern Pennsylvania, and New Jersey (AOU 1983).

The White-throated Sparrow has been only a rare sporadic summer visitor in West Virginia. The first nest of this species was found in Preston County in 1952 (Ganier and Buchanan 1953). Over the years there have been occasional reports of birds in June, but the interpretation of these records is uncertain because this sparrow often delays its northward movement.

The only Atlas records of value were the observations in 1987 of a pair feeding young out of the nest and an adult feeding young in a nest, both records from a bog near the headwaters of Shavers Fork in the so-called Mowrer Tract in Randolph County. The species was not present there in 1988, when the bog was dry, or in 1990 (Bartgis 1992). In 1990, after the close of the Atlas project, a territorial male was found in another bog in the Mowrer Tract (D. Mitchell in. litt.).

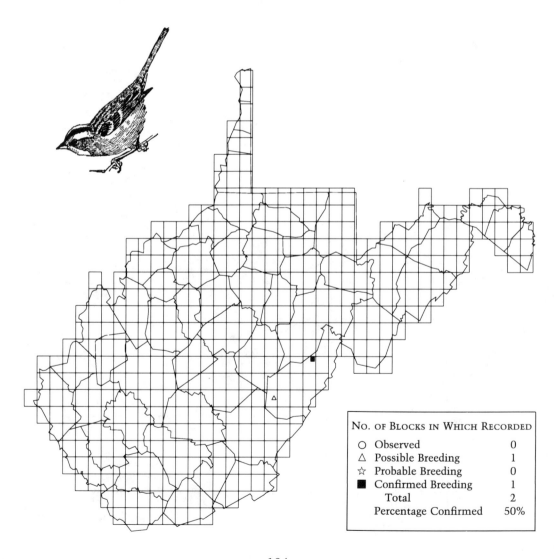

No. of Blocks in Which Recorded	
○ Observed	0
△ Possible Breeding	1
☆ Probable Breeding	0
■ Confirmed Breeding	1
Total	2
Percentage Confirmed	50%

Dark-eyed Junco *Junco hyemalis*

The breeding range of the DARK-EYED JUNCO extends from the limit of trees in the North, south to northern Minnesota, Wisconsin, Pennsylvania, and southern New England, with an extension south along the Appalachian highlands to northern Georgia (AOU 1983). It inhabits open brushy areas, forest edge, and clearings. Populations in the Northeast have shown small declines (Robbins, Bystrak, and Geissler 1986).

The Dark-eyed Junco is a familiar winter bird to most West Virginians, and a drive or a walk along many mountain roads will turn up more of these birds than any other species. It occurs in edge or brushy situations of all forest types. Nests, which are often placed on the banks of road cuts, are easily found. The junco nests only at elevations above 700 to 900 meters (Hall 1983), and the local breeding population generally migrates only to lower elevations.

The Atlas survey confirmed that the junco's main breeding range in West Virginia is in the Allegheny Mountains. It also occurs at higher elevations in the Ridge and Valley Region along the Virginia border, with a few occurrences in the higher parts of Wyoming and Raleigh counties. In Virginia, it nests on the higher mountains along the western border (VSO 1989), and in Kentucky only on Big Black Mountain south of West Virginia (Mengel 1965).

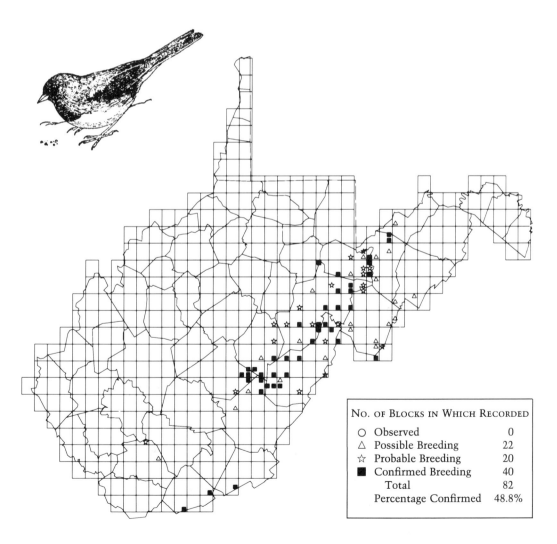

No. of Blocks in Which Recorded	
○ Observed	0
△ Possible Breeding	22
☆ Probable Breeding	20
■ Confirmed Breeding	40
Total	82
Percentage Confirmed	48.8%

Bobolink *Dolichonyx oryzivorus*

The BOBOLINK nests across the northern United States and southern Canada, as far south as northern Missouri, central Indiana, central Ohio, Pennsylvania, and northern New Jersey, with a southward extension into West Virginia. It occupies tall grass areas and deeply cultivated grain fields, as well as clover and alfalfa fields (AOU 1983). Continental populations have declined in this century, probably because many nestings are interrupted by the mowing of hayfields, which is now done earlier in the year than it was in the past.

In West Virginia, the Bobolink nests in open grasslands at high elevations in the Allegheny Mountains, with a few scattered occurrences south to Summers and Raleigh counties. There is also a hilltop population in the extreme northern part of the Northern Panhandle, which is continuous with the Pennsylvania and Ohio populations (Peterjohn and Rice 1991; Brauning 1992).

Although most Atlas records came from high elevations, Bobolinks nested at a low elevation in the South Branch Valley near Petersburg in the summer of 1989.

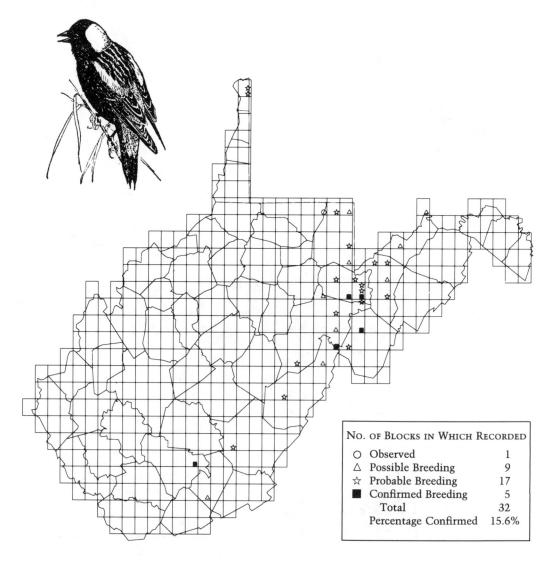

NO. OF BLOCKS IN WHICH RECORDED	
○ Observed	1
△ Possible Breeding	9
☆ Probable Breeding	17
■ Confirmed Breeding	5
Total	32
Percentage Confirmed	15.6%

Red-winged Blackbird *Agelaius phoeniceus*

The RED-WINGED BLACKBIRD is one of the most numerous upland birds in North America. In the first 15 years of the USFWS Breeding Bird Survey, more Red-winged Blackbirds were listed than any other species (Robbins, Bystrak, and Geissler 1986), but in recent years its numbers have been surpassed by the Brown-headed Cowbird. After a surge in population in the 1950s and 1960s, continental populations of Redwings have declined at a rate of 2.8 percent per year (p <0.01) (Robbins, Sauer, Greenberg, and Droege 1989). BBS data reveal that West Virginia populations declined at an annual rate of 3.6 percent (p <0.01) in the period between 1966 and 1989.

The Red-winged Blackbird's breeding range includes the whole continent south of the taiga-tundra ecotone. Usually a bird of the cattail marshes and other wetlands, this species also nests in upland pastures and meadows. In West Virginia, the upland habitat predominates.

Atlas workers found this species in all parts of the state, and the percentage of "confirmed" records was high (55.4%). The few gaps in the Atlas map are probably artifacts of coverage, in that the priority blocks were almost completely forested.

Along with other blackbirds, the Redwing may become an agricultural pest when the large late-summer flocks begin to form. Large concentrations of this species, and other blackbirds, sometimes form in winter. These roosting areas frequently are esthetic nuisances and may indeed be public health hazards.

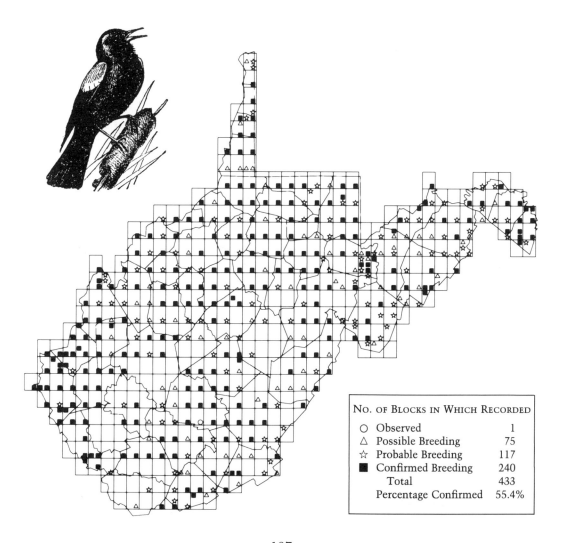

NO. OF BLOCKS IN WHICH RECORDED	
○ Observed	1
△ Possible Breeding	75
☆ Probable Breeding	117
■ Confirmed Breeding	240
Total	433
Percentage Confirmed	55.4%

Eastern Meadowlark *Sturnella magna*

The EASTERN MEADOWLARK was a bird of the midwestern grasslands, but the clearing of the eastern forest produced new habitat for it, and it now occurs throughout the eastern United States. To the west it meets the Western Meadowlark in the Great Plains. The Eastern Meadowlark inhabits grasslands, pastures, and other open lands with adequate grass cover (AOU 1983). Populations in the eastern United States declined in the period from 1966 to 1979 (Robbins, Bystrak, and Geissler 1986).

Hall (1983) speculated that meadowlarks nested in every county in the state, even though definite records were not available from some. Atlas data show that this speculation was incorrect, as the species is apparently absent from parts of southwestern West Virginia. The gaps in the central part of the state could be artifacts resulting from the priority blocks' being entirely forested, but the major gap in the southwest is probably real. The amount of suitable habitat in this area is very limited because of the steep hillsides and small sizes of the grasslands. Strangely, throughout much of the south-central part of the state, seemingly suitable grasslands are nearly birdless.

West Virginia populations of Eastern Meadowlarks declined at a rate of 4.5 percent per year ($p < 0.01$) between 1966 and 1989 (BBS data). Some of this decrease may be due to prolonged cold weather on the bird's wintering grounds in the late 1970s, and some, as with the Bobolink, may be due to changing agricultural practices.

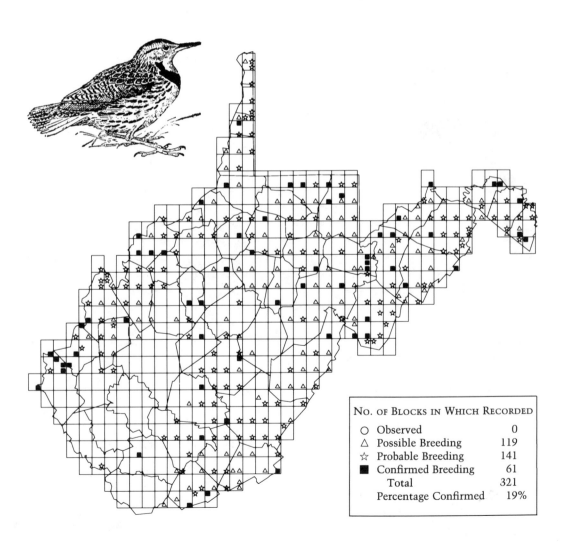

NO. OF BLOCKS IN WHICH RECORDED	
○ Observed	0
△ Possible Breeding	119
☆ Probable Breeding	141
■ Confirmed Breeding	61
Total	321
Percentage Confirmed	19%

Common Grackle *Quiscalus quiscula*

The COMMON GRACKLE is another species that expanded its range and became more common as the original eastern forest was cleared. The grackle now occurs throughout eastern North America south of the tundra, and it is encroaching on the Great Plains to the west. It occurs in partly open situations with scattered trees, such as in forest clearings, along forest edges, and in suburban areas. Despite the range expansions, grackle populations declined by 1.5 percent per year from 1978 to 1987 ($p < 0.01$) (Robbins, Sauer, Greenberg, and Droege 1989).

In West Virginia before the 1950s, grackles were common only in the Eastern Panhandle and in the lower Ohio Valley, but since that time they have spread throughout the state. They remain uncommon in heavily wooded areas but occur wherever there are large openings. The gaps in the central part of the state on the Atlas map may reflect lack of coverage, although the birds may be missing from the steep hills and valleys of the extreme south. Since the beginning of the USFWS Breeding Bird Survey in 1966, West Virginia populations have remained essentially constant.

In their late summer flocks, grackles can become agricultural pests, particularly on corn fields. The increase in range and population has also probably had a negative effect on small-bird populations because of the grackle's nest-robbing tendencies.

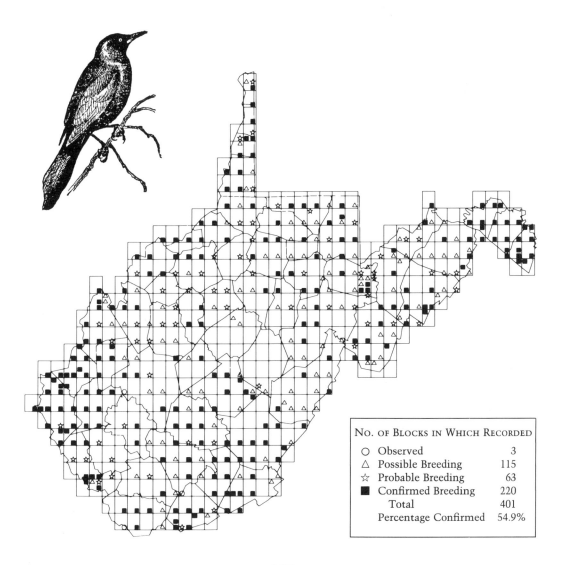

NO. OF BLOCKS IN WHICH RECORDED	
○ Observed	3
△ Possible Breeding	115
☆ Probable Breeding	63
■ Confirmed Breeding	220
Total	401
Percentage Confirmed	54.9%

Brown-headed Cowbird *Molothrus ater*

Originally confined to the open prairies of the West, where it was associated with herds of bison, the BROWN-HEADED COWBIRD has spread to all of eastern North America south of the boreal forest. This species is most numerous in open agricultural land, but it does occur in lightly wooded habitats at all elevations. It is scarce in regions having extensive areas of continuous forest.

The Atlas survey showed that the cowbird occurs throughout the state, but it is rare in the spruce forest and elsewhere at high elevations. West Virginia populations declined at an annual rate of 4.6 percent (p <0.01) in the period from 1966 to 1989 (BBS data).

As the forest becomes fragmented, certain forest-interior birds, such as the Ovenbird, Hooded Warbler, and Wood Thrush, are subjected to increased nest parasitism from the cowbird, greatly reducing the reproductive success of the host species. The cowbird is playing a major part in the population decline of many species of small songbirds.

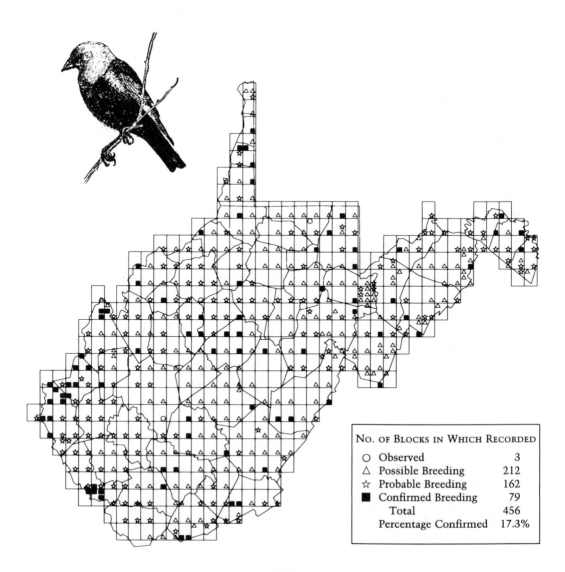

NO. OF BLOCKS IN WHICH RECORDED	
○ Observed	3
△ Possible Breeding	212
☆ Probable Breeding	162
■ Confirmed Breeding	79
Total	456
Percentage Confirmed	17.3%

Orchard Oriole *Icterus spurius*

The ORCHARD ORIOLE is not as conspicuous as its relative the Northern Oriole, nor is it as common. It occurs in eastern North America from central Minnesota, central Wisconsin, and southern Michigan, to southern New York and New England (AOU 1983).

This oriole places its suspended nest high in large trees with spreading canopies in a savannalike situation. Thus it occurs in farmyards and similar places and is uncommon or absent in heavily wooded areas. Continental populations have shown no significant trends in the period from 1978 to 1987 (Robbins, Sauer, Greenberg, and Droege 1989).

West Virginia Atlas workers found the Orchard Oriole to be most prevalent in the Ohio Valley and the western edge of the state, as well as in the Potomac drainage. It was missing or rare in the Allegheny Mountains Region and the higher parts of the Western Hills Region. The steep narrow valleys of the southwestern part of the state have little suitable habitat for this species. Orchard Oriole populations in the state have remained unchanged during the Breeding Bird Survey period.

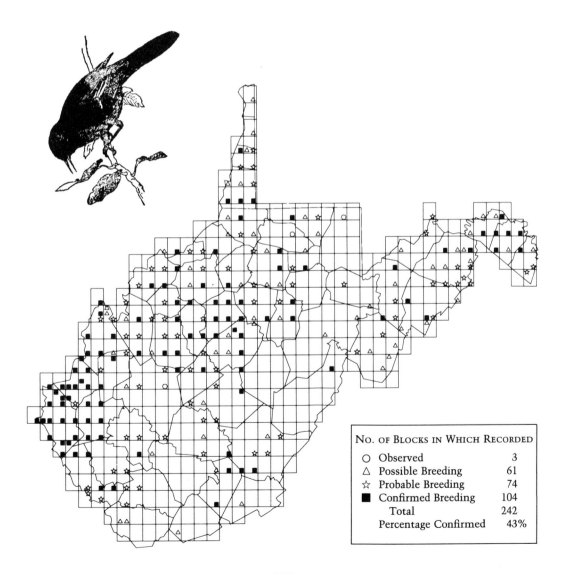

NO. OF BLOCKS IN WHICH RECORDED	
O Observed	3
△ Possible Breeding	61
☆ Probable Breeding	74
■ Confirmed Breeding	104
Total	242
Percentage Confirmed	43%

Northern Oriole *Icterus galbula*

The NORTHERN ORIOLE, more commonly known as the Baltimore Oriole, is a familiar bird to most residents of West Virginia. This oriole's loud and cheerful song and its flashing colors make it a welcome herald of spring. Its continental range embraces almost all of North America east of the Great Plains, south of the boreal forest, and north of the southern pine forest (AOU 1983).

The Northern Oriole places its hanging nest high in large trees with spacious canopies, and it is found wherever such trees are interspersed with open areas. Thus, it is missing from the densely forested regions but occurs widely in farmland, along stream valleys, or in suburban locations. American Elm disease probably deprived this species of many ideal nesting sites. The population in the Northeast experienced an annual decline of 2.9 percent ($p < 0.05$) during the years 1978 to 1987 (Robbins, Sauer, Greenberg, and Droege 1989).

The Atlas project found the Northern Oriole nesting throughout most of the state, confirming Hall's statement to that effect (1983). It is an easy species both to locate and to confirm nesting, so most of the absences from the Atlas map are presumably real, except in some areas in the central part of the state where coverage was not complete. This oriole was missing from the state's heavily forested areas, such as the mountains along the eastern border and the steep slopes and narrow valleys of the hills of the extreme southern section. It also is missing from most of the narrow mountain valleys.

West Virginia populations of Northern Orioles showed a decline of 2.3 percent per year ($p < 0.1$) in the 1966 to 1989 period (BBS data).

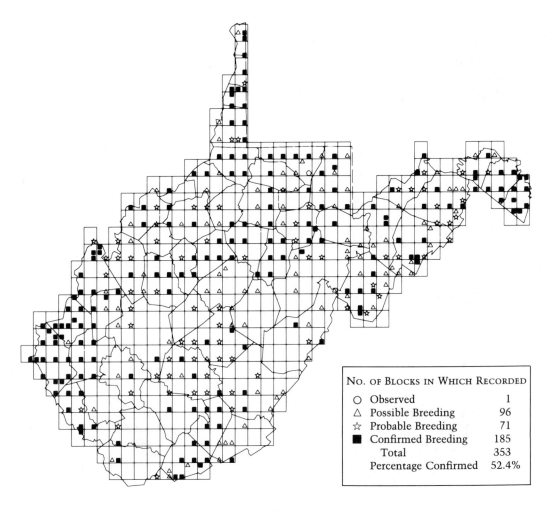

NO. OF BLOCKS IN WHICH RECORDED	
O Observed	1
△ Possible Breeding	96
☆ Probable Breeding	71
■ Confirmed Breeding	185
Total	353
Percentage Confirmed	52.4%

Purple Finch *Carpodacus purpureus*

The PURPLE FINCH is more familiar to West Virginians as a winter species than as a nesting bird; its main breeding range is in the boreal forest of Canada and the mixed forest of the northern states. The range follows the higher Appalachians south as far as West Virginia (AOU 1983).

This finch nests at the edge of the spruce forest or in the middle successional stages of that forest. In recent years, it has been known to nest in planted conifers in the lowlands. The numbers of Purple Finches wintering, and possibly nesting, has been re-duced in recent years by the expansion of the House Finch populations.

The Atlas map shows clearly that the species' main range in West Virginia is along the higher ridges of the Allegheny Mountains, with scattered records in the lowlands. Some of the lowland records may be a consequence of observers' confusing the Purple Finch with the House Finch. *The Ohio Breeding Bird Atlas* (Peterjohn and Rice 1991) shows Purple Finches nesting in areas just to the west of the Northern Panhandle, where the species was reported to be nesting in 1976 (Hall 1983).

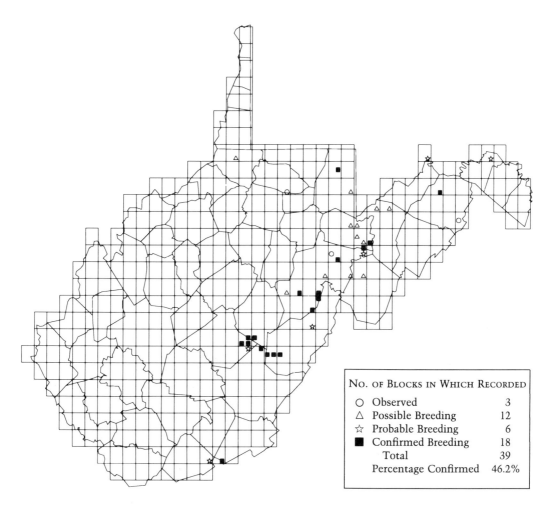

NO. OF BLOCKS IN WHICH RECORDED	
○ Observed	3
△ Possible Breeding	12
☆ Probable Breeding	6
■ Confirmed Breeding	18
Total	39
Percentage Confirmed	46.2%

House Finch *Carpodacus mexicanus*

The HOUSE FINCH is a native of the south-western United States, but a small population was introduced on Long Island, New York, in 1940 (Terres 1980). Since that time, it has thrived and is now found almost throughout the United States east of the Mississippi River. It first appeared in West Virginia in 1971–72, when it was found in the Eastern Panhandle, and it has now spread throughout the state (Hall 1983).

The House Finch occurs commonly in urban gardens, in parks, or on college campuses; in such locations, it nests on buildings or in planted conifers. It is a frequent and abundant patron of winter feeding sta-

tions, but in the summer it is not always quite so evident. The occurrence of House Finches has had a negative effect on the House Sparrow populations in many places, and most people welcome the brightly colored finches as a substitute for the drab sparrow. However, these finches can become objectionable in winter because of their behavior at feeders.

The Atlas data show only a few records in mountain counties and from the center of the state, and in one area in the southwest there are no records at all. This gap is presumably real and not an artifact of lack of coverage, but its cause is unknown.

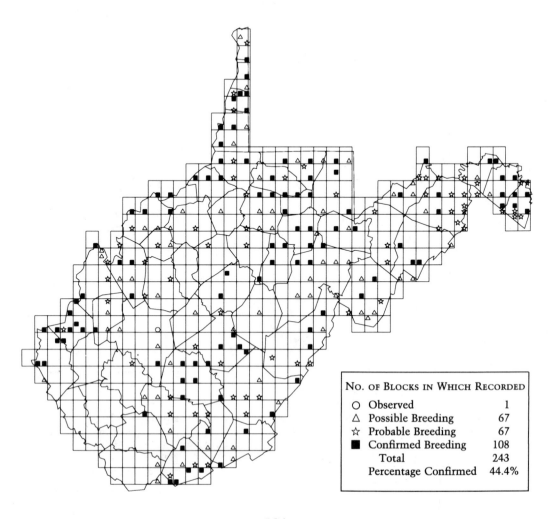

NO. OF BLOCKS IN WHICH RECORDED	
○ Observed	1
△ Possible Breeding	67
☆ Probable Breeding	67
■ Confirmed Breeding	108
Total	243
Percentage Confirmed	44.4%

Pine Siskin *Carduelis pinus*

The PINE SISKIN is a common bird of the Canadian boreal forest. Every winter a few may come as far south as West Virginia, and winter emigrations reach mass proportions on occasion. After some of these great invasions, birds have remained to breed in the spring, far south of their usual range (AOU 1983). Often these birds breed in ornamental conifers in suburban situations, usually near feeders. After raising a brood, they migrate north and may have another nesting in their normal breeding range. There have been a few scattered summer records in the spruce forest area of West Virginia over many years, which suggests that there was a

small breeding population in the higher mountains (Hall 1983). However, prior to the Atlas project, the only confirmed breeding record was at low elevation in Monongalia County (Hall 1983).

During the Atlas project, "confirmed" records of breeding were reported from Monongalia County in the spring of 1988, following the massive 1987 to 1988 invasion, and from Greenbrier County in 1986. The Monongalia County record was of newly fledged young being brought to a feeder. Most of the "possible" records also came in 1988, during which time a nest was also found in nearby Garrett County, Maryland (N. Laitsch pers. com.).

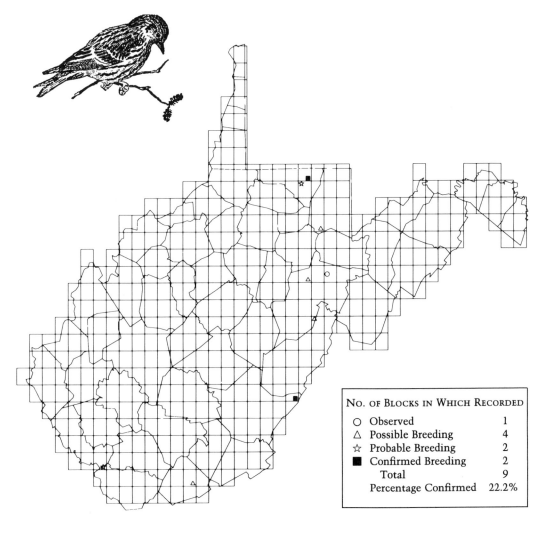

NO. OF BLOCKS IN WHICH RECORDED	
○ Observed	1
△ Possible Breeding	4
☆ Probable Breeding	2
■ Confirmed Breeding	2
Total	9
Percentage Confirmed	22.2%

American Goldfinch *Carduelis tristis*

The AMERICAN GOLDFINCH, often known as the "wild canary," nests throughout the United States except in the extreme south. From 1966 to 1979, the eastern populations underwent a slow decline (Robbins, Bystrak, and Geissler 1986).

The West Virginia Atlas results show this finch nesting throughout the state at all elevations. It usually occurs in brushy open country, where it builds its nest in a variety of small trees (Hall 1983). This species nests in late summer, and young birds are often unfledged in September. This late nesting probably accounts for the low percentage of "confirmed" records of nestings because many Atlas workers had left the field before the American Goldfinch nested.

Since most censuses take place in the early summer, population data for American Goldfinches may not be accurate. The BBS data show a decline in West Virginia populations of 3.4 percent per year ($p < 0.01$) from 1966 to 1989.

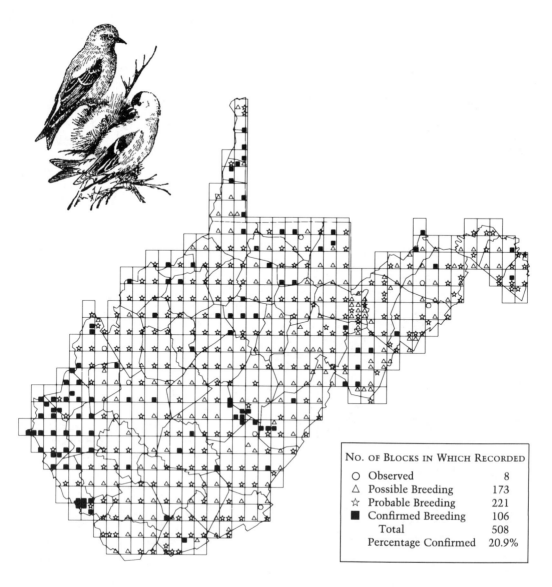

NO. OF BLOCKS IN WHICH RECORDED	
O Observed	8
△ Possible Breeding	173
☆ Probable Breeding	221
■ Confirmed Breeding	106
Total	508
Percentage Confirmed	20.9%

House Sparrow *Passer domesticus*

The HOUSE SPARROW is much maligned and generally ignored by birders. After its first introduction in 1853, it spread rapidly until it now occurs throughout the continent south of the Arctic tree line. Populations started to decline in the early part of this century when the automobile began to replace the horse. They have continued to decline with the changes in farming practices and the decrease in small farms. Numbers have also been reduced by competition with the newly introduced House Finch, which has proved capable of driving the House Sparrow from many places. The BBS data show an annual rate of decrease of 1.4 percent (p <0.01) for the continental populations (Robbins, Sauer, Greenberg, and Droege 1989). West Virginia populations declined at an annual rate of 2.8 percent (p <0.01) from 1966 to 1989 (BBS data).

The Atlas project found the House Sparrow throughout the state, but occurrences were rather scattered in the Allegheny Mountains Region. The bird may nest anywhere around houses or other structures, and the rate of "confirmed" records was correspondingly high (71.9%).

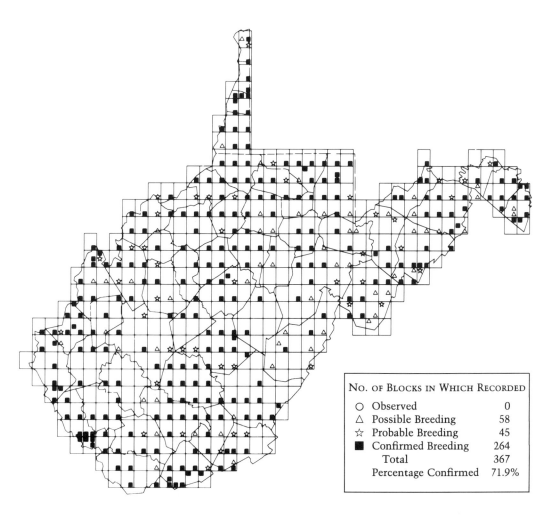

No. of Blocks in Which Recorded	
○ Observed	0
△ Possible Breeding	58
☆ Probable Breeding	45
■ Confirmed Breeding	264
Total	367
Percentage Confirmed	71.9%

Appendixes
Literature Cited
Index

Appendix A
Species of Uncertain Breeding Status
≪ ≫

The following species were encountered during the Atlas project, but observations consisted of only a few "possible" or "observed" records, and no evidence for breeding was confirmed.

Night-Herons

There is one nest record for the BLACK-CROWNED NIGHT-HERON (*Nycticorax nycticorax*) in West Virginia from Wheeling Island in the 1930s (Hall 1983). A few of these nocturnal feeders, which are shy and retiring in their habits, could have been missed by Atlas volunteers. The three scattered Atlas records of adult birds were probably nonbreeding or wandering postbreeding birds. There was a record of a first-year plumaged night-heron near the Bluestone Reservoir, and there were also reports of immature night-herons near Welch and Princeton. These undetermined birds could have been YELLOW-CROWNED NIGHT-HERONS (*Nycticorax violacea*), which breed in Virginia near Pearisburg only a few miles from the West Virginia border (Virginia Atlas data). There were no accepted records for the latter species during the West Virginia Atlas project period.

Green-winged Teal
Anas crecca

The GREEN-WINGED TEAL is a casual summer resident in West Virginia. Its main breeding range is to the north in Canada, but it breeds sporadically south to northern Ohio, Pennsylvania, and northeastern West Virginia. There are two historical breeding records: Tucker County in 1974 and Taylor County in 1971. There was only one Atlas record, a "possible" from Mineral County.

Golden Eagle
Aquila chrysaetos

The GOLDEN EAGLE formerly bred sparsely in the northern Appalachian Mountains. It is now extirpated in New York, once the stronghold of its eastern range (Carroll 1988), and it breeds only in northern Maine (Adamus n.d.) and north in Ontario and Quebec in the East. Its decline is usually attributed to destruction of habitat and shooting.

Although there have been many summer reports from the West Virginia mountains over the years, there are no breeding records for the Golden Eagle from the state (Hall 1983). During the Atlas period, volunteers observed immature and adult Golden Eagles in the Meadow River drainage of Greenbrier County and at Cheat Bridge in Pocahontas County in the summer. Adult Golden Eagles whose breeding attempts have failed and subadults may wander far from the breeding range in summer. Because there are no completely verified records of Golden Eagles nesting in the Appalachians anywhere south of Pennsylvania, and no modern records south of the Adirondacks in New York (Lee and Spofford 1990), presence of birds in the summer in West Virginia should not be considered breeding evidence.

Peregrine Falcon
Falco peregrinus

The accumulated effects of DDT and DDE insecticides drove the PEREGRINE FALCON to the brink of extinction in the late 1960s (Peakall and Kiff 1988). The eastern subspecies, which is now extinct, disappeared from its few scattered eyries in West Virginia by the late 1950s (Hall 1983). The banning of DDT and captive breeding and hacking of peregrines at former eyries in the Northeast have served to reestablish the peregrine as a breeding species in several of these states (Barclay and Cade 1983).

A hacking program was undertaken by the West Virginia Department of Natural Resources Nongame Wildlife Program toward the end of the Atlas project. Seventeen birds were released in the New River Gorge, Fayette County, in the summers of 1987 through 1989, and 24 falcons were released at three sites on North Fork Mountain in Grant and Pendleton counties in 1988 and 1989. Evidence that the Nongame Wildlife Program Peregrine Project is working are the two Peregrine nests found in Grant County in the summer of 1991 after the Atlas period had ended (Stihler 1991). Two chicks hatched from one of these nests in 1991, and three chicks hatched from the same nest in 1992. The eggs in the second nest found in 1991 failed to hatch, and this pair did not nest in 1992. A pair of Peregrines spent the summer of 1992 in Nicholas County but did not nest (Stihler pers. com.).

American Coot
Fulica americana

During the Atlas period, the only record of the AMERICAN COOT was a "possible" report in Hardy County. The American Coot breeds locally from Canada, south to the Gulf of Mexico and Florida. The species seems to have declined since the 1950s, when large flocks were seen on the Ohio River during migration. West Virginia's most recent nesting record was at South Charleston in July 1981 (Hall 1983). There is a Virginia Atlas record near Whitehall, just south of the Jefferson-Berkeley county line (Virginia Atlas data).

Marsh Wren
Cistothorus palustris

The MARSH WREN nests locally in cattail marshes from southern Manitoba, western and southern Ontario, northern Michigan, southwestern Quebec, southern Maine, and southern New Brunswick, south to southern Illinois, central Indiana, northern Ohio, eastern West Virginia, and in coastal marshes south to central Florida, and on the Gulf Coast from central Florida to southern Texas (AOU 1983). This species is limited in distribution to a specialized habitat not sampled well by the USFWS Breeding Bird Survey, so that no trend information is available.

In the past, Marsh Wrens have nested in West Virginia in the Altona Marsh, Jefferson County; the Boaz Marsh, Wood County; and in Tucker County. However, there are no recent records from these stations. An earlier station in Brooke County no longer exists. Atlas workers found "possible" records of breeding of the species in only one block, in Jefferson County.

This species is confined to wetlands of the type that will produce cattail marshes, and the future of the species, nationwide as well as in West Virginia, depends on the continued existence of wetland habitat.

Red Crossbill
Loxia curvirostra

The nomadic RED CROSSBILL fails to obey many of the well-recognized "rules" of bird-life history. Small flocks sometimes move long distances from their normal breeding grounds in the boreal forest and, if they find suitable cone crops in their new location, occasionally nest in late winter far from their original home range (Dickerman 1986). The only acceptable category for "confirmation" of nesting is the finding of an occupied nest. Young birds are fed long after they can fly, and they often do fly long distances from the natal site while still being fed by the parents.

There is a more or less permanent population on Shenandoah Mountain in Virginia just over the West Virginia border, where a nest has been found (Goetz 1981). Many summer records have been reported in West Virginia over the years, but no positive evidence of nesting has ever been obtained (Hall 1983). The Atlas work verified this status with six records, none of which may have been associated with any breeding.

Appendix B
Key to Topographic Maps
<< >>

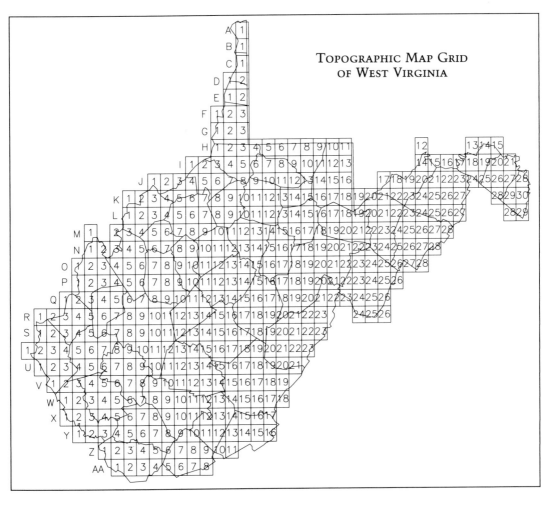

TOPOGRAPHIC MAP GRID
OF WEST VIRGINIA

A-1	East Liverpool South (Ohio)
B-1	Weirton
C-1	Steubenville East (Ohio)
D-1	Tiltonsville (Ohio)
D-2	Bethany
E-1	Wheeling
E-2	Valley Grove
F-1	Businessburg (Ohio)
F-2	Moundsville
F-3	Majorsville
G-1	Powhatan Point
G-2	Glen Easton
G-3	Cameron
H-1	New Martinsville
H-2	Wileyville
H-3	Littleton
H-4	Hundred
H-5	Wadestown
H-6	Blacksville
H-7	Osage
H-8	Morgantown North
H-9	Lake Lynn (Pennsylvania)
H-10	Bruceton Mills
H-11	Brandonville
H-12	Cumberland South (Maryland)
H-13	Bellegrove (Maryland)

H-14	Hancock		K-10	Salem
H-15	Cherry Run (Maryland)		K-11	Wolf Summit
I-1	New Matamoras (Ohio)		K-12	Clarksburg
I-2	Paden City		K-13	Rosemont
I-3	Porters Falls		K-14	Grafton
I-4	Pine Grove		K-15	Thornton
I-5	Big Run		K-16	Fellowsville
I-6	Glover Gap		K-17	Rowlesburg
I-7	Mannington		K-18	Aurora
I-8	Grant Town		K-19	Table Rock (Maryland)
I-9	Rivesville		K-20	Gorman (Maryland)
I-10	Morgantown South		K-21	Mount Storm
I-11	Masontown		K-22	Antioch
I-12	Valley Point		K-23	Burlington
I-13	Cuzzart		K-24	Romney
I-14	Cresaptown		K-25	Augusta
I-15	Patterson Creek (Maryland)		K-26	Hanging Rock
I-16	Oldtown (Maryland)		K-27	Capon Bridge
I-17	Paw Paw		K-28	Inwood
I-18	Great Cacapon		K-29	Middleway
I-19	Stotlers Crossroads		K-30	Charles Town
I-20	Big Pool		L-1	Lubeck
I-21	Hedgesville		L-2	South Parkersburg
J-1	Marietta (Ohio)		L-3	Kanawha
J-2	Belmont		L-4	Petroleum
J-3	Raven Rock (Ohio)		L-5	Cairo
J-4	Bens Run		L-6	Harrisville
J-5	Middlebourne		L-7	Pullman
J-6	Shirley		L-8	Oxford
J-7	Center Point		L-9	New Milton
J-8	Folsom		L-10	Big Isaac
J-9	Wallace		L-11	West Milford
J-10	Shinnston		L-12	Mount Clare
J-11	Fairmont West		L-13	Brownton
J-12	Fairmont East		L-14	Philippi
J-13	Gladesville		L-15	Nestorville
J-14	Newburg		L-16	Colebank
J-15	Kingwood		L-17	Saint George
J-16	Terra Alta		L-18	Lead Mine
J-17	Kitzmiller (Maryland)		L-19	Davis
J-18	Westernport		L-20	Mount Storm
J-19	Keyser		L-21	Greenland Gap
J-20	Headsville		L-22	Medley
J-21	Springfield		L-23	Old Fields
J-22	Levels		L-24	Sector
J-23	Largent		L-25	Rio
J-24	Ridge		L-26	Yellow Spring
J-25	Glengary		L-27	Capon Springs
J-26	Tablers Station		L-28	Berryville (Virginia)
J-27	Martinsburg		L-29	Round Hill
J-28	Shepherdstown		M-1	Pomeroy (Ohio)
K-1	Little Hocking (Ohio)		M-2	Portland (Ohio)
K-2	Parkersburg		M-3	Pond Creek
K-3	Valley Mills		M-4	Rockport
K-4	Willow Island		M-5	Elizabeth
K-5	Schultz		M-6	Girta
K-6	Ellenboro		M-7	Macfarlan
K-7	Pennsboro		M-8	Smithville
K-8	West Union		M-9	Burnt House
K-9	Smithburg		M-10	Auburn

M-11	Vadis	O-15	Walkersville
M-12	Camden	O-16	Rock Cave
M-13	Weston	O-17	Alton
M-14	Berlin	O-18	Cassity
M-15	Century	O-19	Beverly West
M-16	Audra	O-20	Beverly East
M-17	Belington	O-21	Glady
M-18	Montrose	O-22	Whitmer
M-19	Parsons	O-23	Onego
M-20	Mozark Mountain	O-24	Upper Tract
M-21	Blackwater Falls	O-25	Mozer
M-22	Blackbird Knob	O-26	Milam
M-23	Maysville	O-27	Bergton
M-24	Rig	O-28	Orkney Springs (Virginia)
M-25	Moorefield	P-1	Apple Grove
M-26	Needmore	P-2	Arlee
M-27	Baker	P-3	Robertsburg
M-28	Wardensville	P-4	Elmwood
N-1	Cheshire	P-5	Kenna
N-2	New Haven	P-6	Kentuck
N-3	Ravenswood	P-7	Walton
N-4	Sandyville	P-8	Looneyville
N-5	Liverpool	P-9	Triff
N-6	Reedy	P-10	Chloe
N-7	Burning Springs	P-11	Rosedale
N-8	Annamoriah	P-12	Gassaway
N-9	Grantsville	P-13	Sutton
N-10	Tanner	P-14	Newville
N-11	Glenville	P-15	Hacker Valley
N-12	Gilmer	P-16	Goshen
N-13	Peterson	P-17	Pickens
N-14	Roanoke	P-18	Adolph
N-15	Adrian	P-19	Mill Creek
N-16	Buckhannon	P-20	Wildell
N-17	Ellamore	P-21	Sinks of gandy
N-18	Junior	P-22	Spruce Knob
N-19	Elkins	P-23	Circleville
N-20	Bowden	P-24	Franklin
N-21	Harman	P-25	Fort Seybert
N-22	Laneville	P-26	Cow Knob (Virginia)
N-23	Hopeville	Q-1	Athalia (Ohio)
N-24	Petersburg West	Q-2	Glenwood
N-25	Petersburg East	Q-3	Mount Olive
N-26	Lost River State Park	Q-4	Winfield
N-27	Lost City	Q-5	Bancroft
N-28	Wolf Gap	Q-6	Sissonville
O-1	Gallipolis (Ohio)	Q-7	Romance
O-2	Beech Hill	Q-8	Kettle
O-3	Mount Alto	Q-9	Clio
O-4	Cottageville	Q-10	Newton
O-5	Ripley	Q-11	Ivydale
O-6	Gay	Q-12	Strange Creek
O-7	Peniel	Q-13	Herold
O-8	Spencer	Q-14	Little Birch
O-9	Arnoldsburg	Q-15	Erbacon
O-10	Millstone	Q-16	Diana
O-11	Normantown	Q-17	Skelt
O-12	Cedarville	Q-18	Samp
O-13	Burnsville	Q-19	Valley Head
O-14	Orlando	Q-20	Snyder Knob

Q-21	Durbin		T-6	Hager
Q-22	Thornwood		T-7	Griffithsville
Q-23	Snowy Mountain		T-8	Julian
Q-24	Moatstown		T-9	Racine
Q-25	Sugar Grove		T-10	Belle
Q-26	Brandywine		T-11	Cedar Grove
R-1	Catlettsburg (Ohio)		T-12	Montgomery
R-2	Huntington		T-13	Gauley Bridge
R-3	Barboursville		T-14	Ansted
R-4	Milton		T-15	Summersville Dam
R-5	Hurricane		T-16	Mount Nebo
R-6	Scott Depot		T-17	Nettie
R-7	Saint Albans		T-18	Richwood
R-8	Pocatalico		T-19	Fort Mountain
R-9	Big Chimney		T-20	Lobelia
R-10	Blue Creek		T-21	Hillsboro
R-11	Clendenin		T-22	Marlinton
R-12	Elkhurst		T-23	Minnehaha Springs
R-13	Clay		U-1	Louisa (Kentucky)
R-14	Swandale		U-2	Radnor
R-15	Widen		U-3	Kiahsville
R-16	Tiogo		U-4	Ranger
R-17	Cowen		U-5	Big Creek
R-18	Webster Springs		U-6	Mud
R-19	Bergoo		U-7	Madison
R-20	Sharp Knob		U-8	Williams Mountain
R-21	Mingo		U-9	Sylvester
R-22	Cass		U-10	Eskdale
R-23	Green Bank		U-11	Powellton
R-24	Doe Hill (Virginia)		U-12	Beckwith
R-25	Palo Alto		U-13	Fayetteville
R-26	Reddish Knob (Virginia)		U-14	Winona
S-1	Burnaugh		U-15	Corliss
S-2	Lavalette		U-16	Quinwood
S-3	Winslow		U-17	Duo
S-4	West Hamlin		U-18	Trout
S-5	Hamlin		U-19	Droop
S-6	Garretts Bend		U-20	Denmar
S-7	Alum Creek		U-21	Lake Sherwood
S-8	Charleston West		V-1	Webb
S-9	Charleston East		V-2	Wilsondale
S-10	Quick		V-3	Trace
S-11	Mammoth		V-4	Chapmanville
S-12	Bentree		V-5	Henlawson
S-13	Lockwood		V-6	Clothier
S-14	Gilboa		V-7	Wharton
S-15	Summersville		V-8	Whitesville
S-16	Craigsville		V-9	Dorothy
S-17	Camden on Gauley		V-10	Pax
S-18	Webster Springs Southwest		V-11	Oak Hill
S-19	Webster Springs Southeast		V-12	Thurmond
S-20	Woodrow		V-13	Danese
S-21	Edray		V-14	Rainelle
S-22	Clover Lick		V-15	Rupert
S-23	Paddy Knob		V-16	Cornstalk
T-1	Fallsburg (Kentucky)		V-17	Williamsburg
T-2	Prichard		V-18	Anthony
T-3	Wayne		V-19	Alvon
T-4	Nestlow		W-1	Naugatuck
T-5	Branchland		W-2	Myrtle

W-3	Holden	Y-2	Majestic
W-4	Logan	Y-3	Wharncliffe
W-5	Amherstdale	Y-4	Gilbert
W-6	Lorado	Y-5	Baileysville
W-7	Pilot Knob	Y-6	Pineville
W-8	Arnett	Y-7	Mullens
W-9	Eccles	Y-8	Rhodell
W-10	Beckley	Y-9	Odd
W-11	Prince	Y-10	Flat Top
W-12	Meadow Creek	Y-11	Pipestem
W-13	Meadow Bridge	Y-12	Forest Hill
W-14	Dawson	Y-13	Greenville
W-15	Asbury	Y-14	Union
W-16	Lewisburg	Y-15	Gap Mills
W-17	White Sulphur Springs	Y-16	Paint Bank
W-18	Jerrys Run (Virginia)	Z-1	Panther
X-1	Williamson	Z-2	Iaeger
X-2	Delbarton	Z-3	Davy
X-3	Barnabus	Z-4	Welch
X-4	Man	Z-5	Keystone
X-5	Mallory	Z-6	Crumpler
X-6	Oceana	Z-7	Matoaka
X-7	Matheny	Z-8	Athens
X-8	McGraws	Z-9	Lerona
X-9	Lester	Z-10	Peterstown
X-10	Crab Orchard	Z-11	Lindside
X-11	Shady Spring	AA-1	Bradshaw
X-12	Hinton	AA-2	War
X-13	Talcott	AA-3	Gary
X-14	Alderson	AA-4	Anawalt
X-15	Fort Spring	AA-5	Bramwell
X-16	Ronceverte	AA-6	Bluefield
X-17	Glace	AA-7	Princeton
Y-1	Matewan	AA-8	Oakvale

Literature Cited

<< >>

Adamus, P. R. n.d. *Atlas of breeding birds in Maine, 1978–1983.* Augusta: Maine Department of Inland Fisheries and Wildlife.

American Ornithologists' Union. 1957. *Check-list of North American birds.* 5th ed. Baltimore: Lord Baltimore Press.

———. 1983. *Check-list of North American birds.* 6th ed. Lawrence, Kans.: Allen Press.

Barclay, J. H., and T. J. Cade. 1983. Restoration of the Peregrine Falcon in the eastern United States. *Bird Conservation* 1: 3–37.

Bartgis, R. L. 1992. Southernmost breeding record of the White-throated Sparrow in Appalachia. *Redstart* 59: 80–81.

Beatty, J. 1988. Eastern Screech-Owl: Population dynamics and potential predation upon the meadow vole. *Redstart* 55: 98–104.

Bednarz, J. C., and J. J. Dinsmore. 1981. Status, habitat use, and management of Red-shouldered Hawks in Iowa. *Journal of Wildlife Management* 45: 236–41.

Bellrose, F. C., and R. A. Heister. 1987. The Wood Duck. In *Audubon Wildlife Report 1987*, edited by R. L. DiSilvestro. New York: Academic Press.

Bent, A. C. 1926. *Life histories of North American marsh birds*, U.S. National Museum Bulletin, no. 135, Washington, D.C. Dover ed., 1963.

Bonney, R. E., Jr. 1988. Worm-eating Warbler. In *The atlas of breeding birds in New York State*, edited by R. F. Andrle and J. R. Carroll, pp. 404–05. Ithaca, N.Y.: Cornell University Press.

Brauning, D. W., ed. 1992. *Atlas of breeding birds in Pennsylvania.* Pittsburgh: University of Pittsburgh Press.

Brewer, R. 1963. Reproductive relationships of Black-capped and Carolina chickadees. *Auk* 80: 9–47.

Brooks, M. 1938a. Bachman's Sparrow in the north-central portion of its range. *Wilson Bulletin* 50: 86–109.

———. 1938b. The Eastern Lark Sparrow in the upper Ohio Valley. *Cardinal* 4: 181–200.

———. 1944. *Checklist of West Virginia birds.* West Virginia Agricultural Experiment Station Bulletin no. 316.

Buckelew, A. R., Jr. 1976. Birds of the West Virginia Northern Panhandle. *Redstart* 43: 90–107.

———. 1991. Recent Northern Goshawk breeding records from the West Virginia highlands. *Redstart* 58: 74–75.

Carroll, J. R. 1988. Golden Eagle. In *The atlas of breeding birds in New York State*, edited by R. F. Andrle and J. R. Carroll, pp. 116–17. Ithaca, N.Y.: Cornell University Press.

Connor, P. F. 1988. Pied-bill Grebe. In *The atlas of breeding birds in New York State*, edited by R. F. Andrle and J. R. Carroll, pp. 28–29. Ithaca, N.Y.: Cornell University Press.

Craighead, J. J., and F. C. Craighead, Jr. 1956. *Hawks, owls and wildlife.* Dover ed. Harrisburg, Pa.: Stackpole Co., 1969.

Crum, J. M. 1989. Spruce Knob Ruffed Grouse Management Area population trends. In *1989 Small Game Bulletin*, pp. 22–24. West Virginia Department of Natural Resources, Wildlife Division.

Dickerman, R. W. 1986. A review of the Red Crossbill in New York State. Part 1. Historical and nomenclatural background. *Kingbird* 36: 73–78.

DiGiovanni, D. M. 1990. *Forest statistics for West Virginia—1975 and 1989.* USDA Forest Service, Northeastern Forest Experiment Station Resource Bulletin NE-114.

Dunning, J. B., Jr., and B. D. Watts. 1990. Regional differences in habitat occupancy by Bachman's Sparrow. *Auk* 107: 463–72.

Eddy, G. A. 1988. Yellow-rumped Warbler nesting record in West Virginia. *Redstart* 55: 56–57.

Edeburn, R. M. 1968. Breeding range extension of Saw-whet Owls in West Virginia. *Wilson Bulletin* 80: 232.

Ehrlich, P. R., D. S. Dobkin, and Wheye, D. 1988. *The birder's handbook.* New York: Simon and Schuster.

Federal Register. January 6, 1989.

Forbes, J. E., and D. W. Warner. 1974. Behavior of a radio-tagged Saw-whet Owl. *Auk* 91: 783–95.

Fye, W. L. 1984. Nesting Upland Sandpiper in Clarion County, Pa. *Redstart* 51: 68.

Ganier, A. F., and F. W. Buchanan. 1953. Nesting of White-throated Sparrow in West Virginia. *Wilson Bulletin* 65: 277–79.

Gill, F. B. 1985. Birds. In *Species of special concern in Pennsylvania*, edited by H. H. Genoways and F. J. Brenner, pp. 97–351. Pittsburgh: Carnegie Museum of Natural History, Special Publication no. 11.

Goetz, W. J. 1981. Nesting Red Crossbills in Rockingham County, Virginia. *Redstart* 48: 90–92.

Green, N. 1985. The Bald Eagle. In *Audubon wild-*

Literature Cited

life report, 1985, edited by R. L. DiSilvestro, pp. 508–31. New York: National Audubon Society.

Gross, A. O. 1942. Cliff Swallows. In *Life histories of North American flycatchers, larks, swallows, and their allies*, edited by A. C. Bent, pp. 463–84. U.S. National Museum Bulletin no. 195. Dover ed. Washington, D.C.: Smithsonian Institution, 1963.

Hall, G. A. 1969. Breeding range expansion of the Brown Creeper in the Middle Atlantic States. *Redstart* 36: 98–103.

———. 1983. *West Virginia birds*. Pittsburgh: Carnegie Museum of Natural History, Special Publication no. 7.

———. 1984. Population decline of neotropical migrants in an Appalachian Forest. *American Birds* 38: 14–18.

———. 1985. The Appalachian Region. *American Birds* 39: 911–14.

———. 1988. The Appalachian Region. *American Birds* 42: 1286–89.

———. 1989. The Appalachian Region. *American Birds* 43: 1314–17.

———. 1991. The Appalachian Region. *American Birds* 45: 1114–17

Handley, C. O., Sr. 1976. *Birds of the Great Kanawha Valley*. Parsons, W. Va.: McLain Publishing.

Harrison, H. H. 1975. *A field guide to birds' nests*. Boston: Houghton Mifflin.

Hicks, L. E. 1935. *Distribution of the breeding birds of Ohio*. Ohio Biological Survey, Bulletin no. 32.

Hurley, G. F. 1972. Swainson's Warbler distribution in West Virginia. *Redstart* 39: 110–12.

Igou, T. D. 1984. Further observations of Common Moorhens at McClintic. *Redstart* 51: 140.

———. 1986. Nesting Cliff Swallows at Beech Fork State Park. *Redstart* 53: 137.

Johnsgard, P. A. 1975. *Waterfowl of North America*. Bloomington, Ind.: Indiana University Press.

Johnston, D. W. 1971. Ecological aspects of hybridizing chickadees *(Parus)* in Virginia. *American Midland Naturalist* 85: 124–34.

Kiff, L., T. D. Igou, H. E. Slack III, and L. Wilson. 1986. *Birds of the lower Ohio River Valley in West Virginia*. Wheeling, W. Va.: Brooks Bird Club, Special Publication no. 1.

Kletzly, R. C. 1976. *American Woodcock in West Virginia*. West Virginia Department of Natural Resources, Bulletin no. 8.

Krebs, J. R. 1974. Colonial nesting and social feeding as strategies for exploiting food resources in the Great Blue Heron *(Ardea herodias)*. *Behavior* 51: 99–131.

Laitsch, N. 1983. The 1982 Foray bird list. *Redstart* 50: 4–12.

———. 1988. Pers. com. 1203 East Park Boulevard, East Liverpool, Ohio 43920.

Laughlin, S. B., and D. P. Kibbe. 1985. *The atlas of breeding birds of Vermont*. Hanover, N.H., and London: University Press of New England.

Lee, D. S., and W. R. Spofford. 1990. Nesting of Golden Eagles in the central and southern Appalachians. *Wilson Bulletin* 102: 693–98.

Legg, W. 1942. Swainson's Warbler in Nicholas County, West Virginia. *Wilson Bulletin* 54: 252.

Levine, E. 1988. Yellow-bellied Sapsucker. In *The atlas of breeding birds in New York State*, edited by R. F. Andrle and J. R. Carroll, pp. 230–31. Ithaca, N.Y.: Cornell University Press.

Marshall, J. T. 1988. Birds lost from a giant sequoia forest during the last fifty years. *Condor* 90: 359–72.

Meade, G. M. 1988. Chuck-will's-widow. In *The atlas of breeding birds in New York State*, edited by R. F. Andrle and J. R. Carroll, pp. 216–17. Ithaca, N.Y.: Cornell University Press.

Mengel, R. M. 1965. *The birds of Kentucky*. American Ornithologists' Union, Ornithological Monograph no. 103.

Michael, E. D. 1990. The snipe and other wildlife of Canaan Valley wetlands. *Wonderful West Virginia* 54: 25–26.

Mitchell, D. 1989. Black Vulture nest in Grant County, West Virginia. *Redstart* 56: 84.

Monroe, B. L., Jr. 1988. Summary of highest counts of individuals for Canada and the United States. *American Birds* 42: 1184–90.

———. 1989. Summary of highest counts of individuals for Canada and the United States. *American Birds* 43: 1213–20.

Pack, J. C. 1989. Wild Turkey. *1989 Big Game Bulletin*. West Virginia Department of Natural Resources, Wildlife Division.

Peakall D. B., and L. F. Kiff. 1988. DDE contamination in Peregrine and American kestrels and its effect on reproduction. In *Peregrine Falcon populations: Their management and recovery*, edited by T. J. Cade, J. H. Enderson, C. G. Thelander, and C. M. White, pp. 337–50. Boise, Idaho: The Peregrine Fund.

Peterjohn, B. G. 1989. *The birds of Ohio*. Bloomington, Ind.: Indiana University Press.

Peterjohn, B. G., and D. L. Rice. 1991. *The Ohio breeding bird atlas*. Columbus, Ohio: Ohio Department of Natural Resources.

Peterson, R. T. 1980. *A field guide to the birds*. 4th ed. Boston: Houghton Mifflin.

Rieffenberger, J. C. 1988. Great Blue Herons nesting on Cheat Mountain. *Redstart* 55: 104–05.

Rieffenberger, J. C., and J. M. Crum. 1989. Ruffed Grouse. In *1989 Small Game Bulletin*. West Virginia Department of Natural Resources, Wildlife Division.

Robbins, C. S., D. Bystrak, and P. H. Geissler. 1986. *The Breeding Bird Survey: Its first fifteen years, 1965–1979*. Resource Publication No. 157. Washington, D.C.: U.S. Department of the Interior. Fish and Wildlife Service.

Robbins, C. S., D. K. Dawson, and B. A. Dowell. 1989. Habitat area requirements of breeding forest birds

Literature Cited

of the Middle Atlantic states. *Wildlife Monographs* 104: 1–34.

Robbins, C. S., J. R. Sauer, R. S. Greenberg, and S. Droege. 1989. Population declines in North American birds that migrate to the neotropics. *Proceedings of the National Academy of Sciences.* U.S.A. 86: 7658–62.

Robbins, C. S., J. W. Fitzpatrick, and P. B. Hamel. 1992. A warbler in trouble: *Dendroica cerulea*. In *Ecology and conservation of Neotropical migrant landbirds,* edited by J. M. Hagan, III and D. Johnston, pp. 549–61. Washington, D.C.: Smithsonian Institution.

Robinson, S. K. 1992. Population dynamics of breeding birds in a fragmented Illinois landscape. In *Ecology and conservation of Neotropical migrant landbirds,* edited by J. M. Hagan, III and D. Johnston, pp. 408–18. Washington, D.C.: Smithsonian Institution.

Samuel, D. E. 1969. House Sparrow occupancy of Cliff Swallow nests. *Wilson Bulletin* 81: 103–04.

———. 1971. The breeding biology of Barn and Cliff swallows in West Virginia. *Wilson Bulletin* 83: 284–301.

Sandercox, A., and A. R. Buckelew, Jr. 1983. An unusual nest site and diet of Common Barn-Owls in Bethany, West Virginia. *Redstart* 50: 82–84.

Sharrock, J.T.R. 1976. *The atlas of breeding birds in Britain and Ireland.* Tring, Herts., G.B.: British Trust for Ornithology.

Shreve, A. 1977. Territorial Short-billed Marsh Wren in Kanawha County. *Redstart* 44: 94–97.

Smith, B., and G. A. Hall. 1983. Census No. 82. Stunted spruce-shrub community. *American Birds* 37: 76.

Snyder, N.F.R., H. A. Snyder, J. L. Linar, and R. T. Reynolds. 1973. Organochlorines, heavy metals, and the biology of North American accipiters. *BioScience* 23: 300–05.

Stewart, P. A. 1978. Behavioral interactions and niche separation in Black and Turkey vultures. *Living Bird* 17: 79–84.

Stihler, C. 1991. Peregrine Falcons nest in West Virginia. *Redstart* 58: 110–12.

———. 1992. Pers. com. West Virginia DNR, P.O. Box 67, Elkins, W. Va. 26241.

Strausbaugh, P. D., and E. L. Core. 1970–77. *Flora of West Virginia.* 2nd ed. Grantsville, W. Va.: Seneca Books. (Now available in a bound printing from West Virginia University Press.)

Sutton, G. M. 1923. Notes on the nesting of the Wilson's Snipe in Crawford County, Pennsylvania. *Wilson Bulletin* 35: 191–202.

———. 1928. Extension of the breeding range of the Turkey Vulture in Pennsylvania. *Auk* 45: 501–03.

Tanner, J. T. 1952. Black-capped and Carolina chickadees in the southern Appalachian Mountains. *Auk* 69: 407–24.

Tate, J., Jr. 1986. The Blue List for 1986. *American Birds* 40: 227–36.

Tate, J., Jr., and D. J. Tate. 1982. The Blue List for 1982. *American Birds* 36: 126–35.

Terres, J. K. 1980. *The Audubon Society encyclopedia of North American birds.* New York: Albert A. Knopf.

Todd, W.E.C. 1940. *Birds of western Pennsylvania.* Pittsburgh: University of Pittsburgh Press.

Virginia Society of Ornithology. 1989. *Virginia's breeding birds: An atlas workbook.* Richmond, Va.: Virginia Society of Ornithology.

Ward, J. 1987. The 1986 Foray bird list. *Redstart* 54: 3–11.

Ward, R., and D. A. Ward. 1974. Songs in contiguous populations of Black-capped and Carolina chickadees in Pennsylvania. *Wilson Bulletin* 86: 344–56.

West Virginia Department of Natural Resources. n.d. *Vertebrate species of concern in West Virginia.* Elkins, W. Va.

Whitmore, R. C., and G. A. Hall. 1978. The response of passerine species to a new resource: Reclaimed surface mines in West Virginia. *American Birds* 32: 6–9.

Worthington, G. 1989. First confirmed nesting of Yellow-bellied Flycatchers in West Virginia. *Redstart* 56: 58–60.

Index

<< >>

Index